Mammalian Neuroendocrinology

GEOFFREY W. BENNETT,

Lecturer in Physiology and Pharmacology,
University of Nottingham Medical School

SAFFRON A. WHITEHEAD,

Lecturer in Physiology,
St George's Hospital Medical School, London

CROOM HELM
London & Canberra

Croom Helm Ltd, Provident House, Burrell Row,
Beckenham, Kent BR3 1AT

British Library Cataloguing in Publication Data

Bennett, Geoffrey W.
Mammalian neuroendocrinology.
1. Neurophysiology 2. Neuroendocrinology
I. Title
599'.0142 QP356.4

ISBN 0-7099-0638-2
ISBN 0-7099-0674-9 Pbk

Printed and bound in Great Britain
by Billing and Sons Ltd, Worcester.

CONTENTS

PREFACE

Neuroendocrinology is no longer simply a branch of endocrinology but has become an established area of scientific endeavour which is of growing interest to physiologists, biochemists, pharmacologists and zoologists alike. This new status has arisen largely as a result of recent research advances which followed the discovery of active brain peptides and has markedly advanced our knowledge of the central control mechanisms of neural and endocrine systems. This book has been written as an introduction to mammalian neuroendocrinology and provides an overview of this highly complex field with particular emphasis on the hypothalamus and the control of pituitary functions. It is suitable for science undergraduate or clinical students and graduate or research scientists interested in broadening their interests into this rapidly expanding field.

The first chapter summarises the basic facts and principles of neuroendocrinology and provides the fundamental information relevant to the reader who is completely new to the subject. It outlines the anatomy of the hypothalamic-pituitary axis, the control mechanisms of pituitary secretions and the feedback and hormone receptor interactions which are well established and understood. The second chapter describes the main methods employed in neuroendocrine research so that the reader has some insight into both well-established (classical) and more recent techniques, which include histological, physiological, biochemical, neurochemical and pharmacological methodology. Subsequent chapters then cover various areas in greater depth and incorporate details of the neuroendocrine control of the posterior pituitary, the intermediate lobe of the pituitary and the control of hormone secretions from the anterior pituitary gland. The later chapters are concerned with aspects of particular importance or interest to neuroendocrinology and cover topics such as circadian rhythms of hormone secretion, seasonal breeding and hormones and behaviour. Also included is a specific chapter on brain neuropeptides which illustrates the large amount of information already available on the nature and distribution of this recently discovered family of messenger substances and their relevance to neuroendocrine control mechanisms. The final chapter outlines important clinical aspects which include disorders of the hypothalamic pituitary axis in humans and the emerging diagnostic and therapeutic value of hypothalamic peptide hormones and related drugs.

The book contains well over 150 references and these are listed at the end of the appropriate chapters. Generally the references quoted are recent reviews from journals and books from which the reader can obtain further information and more specific references to original papers. In addition references are given to closely related topics not covered by the present book: such subjects include hormones and neuropeptides of the gastrointestinal tract and evolution of hormones. A further list of recommended books, not listed in the text, is given before the index. The book is divided into subsections and use has been made of cross referencing within the book, so that the reader is directed to related information given in other sections or chapters.

We are deeply grateful to Nicola Barrett for drawing the diagrams presented in the book, to many of our colleagues and friends for helpful advice and criticism in the preparation of the manuscript and particularly to our respective spouses for their patience and understanding throughout.

ABBREVIATIONS

Anterior Pituitary Hormones

ACTH	Adrenocorticotrophic hormone (corticotrophin)
LPH	Lipotrophic hormone (lipotrophin)
LH	Luteinising hormone
FSH	Follicle stimulating hormone
TSH	Thyroid stimulating hormone
PRL	Prolactin
GH	Growth hormone (somatotrophin)

Intermediate Lobe Hormones

MSH	Melanocyte stimulating hormone
CLIP	Corticotrophin-like intermediate lobe peptide

Posterior Pituitary Hormones

VP	Vasopressin (antidiuretic hormone, ADH)
OXY	Oxytocin
NP	Neurophysin

Hypothalamic Hormones and Factors

CRF	Corticotrophin releasing factor
Gn-RH	Gonadotrophin releasing hormone (luteinising hormone releasing hormone, LHRH)
PIF	Prolactin release inhibiting factor
PRF	Prolactin releasing factor
TRH	Thyrotrophin releasing hormone
SOM	Somatostatin
GHRF	Growth hormone releasing factor
MSH-IF	MSH release inhibiting factor
MSH-RF	MSH releasing factor

Abbreviations

Neurotransmitters

ACh	Acetylcholine
DA	Dopamine
NA	Noradrenaline
A	Adrenaline
5-HT	5-hydroxytryptamine
HIS	Histamine
GABA	γ-Aminobutyric Acid

Amino Acids (present in peptide hormones and neuropeptides)

Ala	Alanine
Arg	Arginine
Asn	Asparagine
Asp	Aspartic Acid
Cys	Cysteine
Glu	Glutamic Acid
Gln	Glutamine
Gly	Glycine
His	Histidine
Ile	Isoleucine
Leu	Leucine
Lys	Lysine
Met	Methionine
Phe	Phenylalanine
Pro	Proline
Ser	Serine
Thr	Threonine
Trp	Tryptophan
Tyr	Tyrosine
Val	Valine

1

BASIC PRINCIPLES OF NEUROENDOCRINOLOGY

Neuroendocrinology may be defined as the study of the physiological processes involved in integrating neural and endocrine functions. This chapter will summarise our present understanding of the mammalian tissues and physiological processes which are the concern of neuroendocrinology and will introduce the ways in which cells of the nervous and endocrine systems respond to environmental changes and to each other to regulate bodily functions by the mediation of specific chemical agents.

Subsequent chapters will elaborate in more detail the complex neuroendocrine roles played by the hypothalamus and pituitary in particular, and final chapters will cover some important developments which have emerged from recent neuro-endocrine research, including some novel clinical applications for the diagnosis and treatment of neuroendocrine disorders.

1:1 Neurotransmitters, Neurohormones and Hormones

Chemical agents which are released from nerve cells (neurones) and act directly on neighbouring neurones or other cells are termed neurotransmitters, whereas chemicals released from endocrine cells directly into the blood circulation and transported to distant target cells are called hormones (Figure 1.1). As a result of histochemical studies pioneered in the 1940s it emerged that certain nerve cells are neurosecretory in that they release chemical agents into the blood.

Figure 1.1: Types of chemical communication between cells. M — chemical messenger (neurotransmitters, neurohormones or hormones).

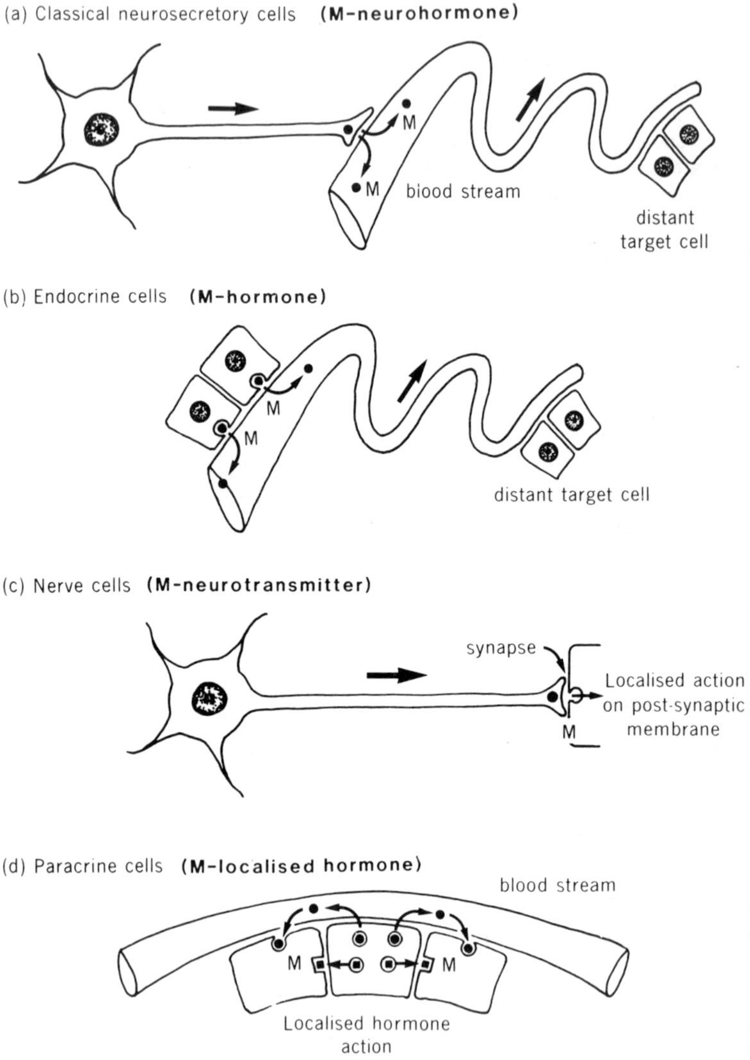

Table 1.1: Main Types of Chemical Messenger

CHEMICAL MESSENGER	SITES OF RELEASE
1. *Peptides* (Hormones, Neurohormones, Possible Neurotransmitters) Chains of amino acids of length 2 to several hundred joined by peptide bonds. These consist of any of 20 natural amino acids.	Many endocrine glands, especially Pituitary. Hypothalamus, Central and Peripheral Nervous Systems.
2. *Modified amino acids*	
(a) Thyroid hormones (Hormones) Formed from mono- and di-iodinated tyrosine (see Figure 6.1)	Thyroid gland.
(b) Catecholamines (Hormones, Neurotransmitters) Adrenaline, Noradrenaline andDopamine formed from tyrosine via L-dopa (see [10]).	Adrenal Medulla, Central and Peripheral Nervous Systems.
(c) Indoleamines (Neurotransmitters) Mainly 5-hydroxytryptamine formed from tryptophan (see [10]).	Central Nervous System.
3. *Amino acids* (Possible Neurotransmitters) Certain acidic amino acids e.g. glutamate, aspartate and γ-amino butyric acid (GABA), and glycine.	Central Nervous System.
4. *Steroids* (Hormones) Formed from cholesterol and have a characteristic planar structure of carbon atoms (see Figure 5.1)	Adrenal cortex and Gonads.
5. *Prostaglandins* (Local Hormones) Formed from arachidonic acid with characteristic structure (see [17])	Many Reproductive, Neural and Other Tissues.

Consequently the term neurohormone was introduced and the science of neuroendocrinology was born.

The wealth of neuroendocrine research which has been undertaken, particularly during the last 20 years, has shown that a minute region of the mammalian forebrain, called the hypothalamus, is an important region of neurohormone production and that the pituitary gland, situated directly beneath, acts as the principal target organ or mediator for the action of these neurohormones. More profoundly perhaps, recent research has further demonstrated previously unforeseen similarities and relationships between nervous and endocrine systems. What has become abundantly clear is that there are chemical messengers, very often small peptides in nature, which are produced and released by nerve cells in the hypothalamus and other regions of the central and peripheral nervous system that are identical with those produced by cells in the pituitary and peripheral endocrine organs. Thus the distinction between neurotransmitter, neurohormone and hormone may not always be obvious on the basis of chemical identity alone, and the same substance may possess different functions in different body tissues (Figure 1.1 and Table 1.1).

It should be mentioned in passing that a further type of localised chemical communication between cells occurs in some tissues. In this case the chemical messenger is not transported any distance in blood and only acts in a highly circumscribed area, the secretory cells are not true endocrine cells and have been termed paracrine cells (Figure 1.4 and [4]).

1:2 The Hypothalamus

The hypothalamus may be defined in either an anatomical or physiological sense. Anatomically, the hypothalamus is an area of grey matter in the basal part of the forebrain. It is divided into two symmetrical halves by the third ventricle – one of the chambers of brain filled with cerebrospinal fluid (CSF). The hypothalamus is bound rostrally (towards the nose) by the optic chiasma, caudally (towards the tail) by the mamillary bodies, laterally by the optic tracts of the eyes and dorso/laterally by the brain region called the thalamus.

The principal nuclei (groups of nerve cell bodies) of the hypothalamus are bilaterally arranged around the third ventricle. Those concerned with pituitary regulation tend to a medial distribution, whilst those of the lateral hypothalamus are involved in other functions such as regulation of food intake. There are species differences in the presence and distribution of the hypothalamic nuclei and in the arrangement of the hypothalamic–hypophyseal tissues. This can be seen for example, by comparing the human situation with that of the most popular experimental animal, the rat (Figures 1.2a and b). It should be pointed out that these diagrams, although apparently detailed, are only a simple representation of the complexity of structures involved. For the purposes of precise experimental operations or manipulations on individual hypothalamic nuclei detailed atlases of brain nuclei for a number of species (including rat, guinea pig, cat, sheep, monkey, man) are available.

Physiologically, the hypothalamus is an important centre in the regulation of the autonomic nervous system. Visceral afferent fibres have been shown to enter the hypothalamus and efferent fibres carry impulses via subsidiary centres in the brain stem down the spinal cord

Figure 1.2: Distribution of Principal Nuclei (N) in (a) Rat and (b) Human Hypothalamus.

Abbreviations (N = nucleus)

Anterior Hypothalamus	Preoptic area	POA
	Anterior hypothalamic area (rat)	AHA
	Supraoptic N.	SO
	Suprachiasmatic N.	SC
	Paraventricular N.	PV
Mid Hypothalamus	Ventromedial N.	VM
	Dorsomedial N.	DM
	Lateral tuberal N. (human)	LT
	Lateral periventricular N. (human)	LP
	Infundibular N. (human)	IF
	Arcuate N. (rat)	ARC
Posterior Hypothalamus	Posterior hypothalamic N.	PH
	Premamillary N.	PM
	Tuberomamillary N. (human)	TM
	Supramamillary N. (human)	SM
	Supramamillary area (rat)	SMA
	Medialmamillary N. (rat)	MM
	Posteriormamillary N. (rat)	PMT
	Lateral hypothalamic area	LHA
Other landmarks	Anterior commissure	AC
	Mamillary body	MB
	Optic chiasma	OC
	Third ventricle	IIIV
	Infundibulum	I
	Fornix	FX
	Medial forebrain bundle	MFB
Pituitary	Neurohypophysis	N
	Adenohypophysis	A
	Pars intermedia	PI
	Pars tuberalis	PT

(a)

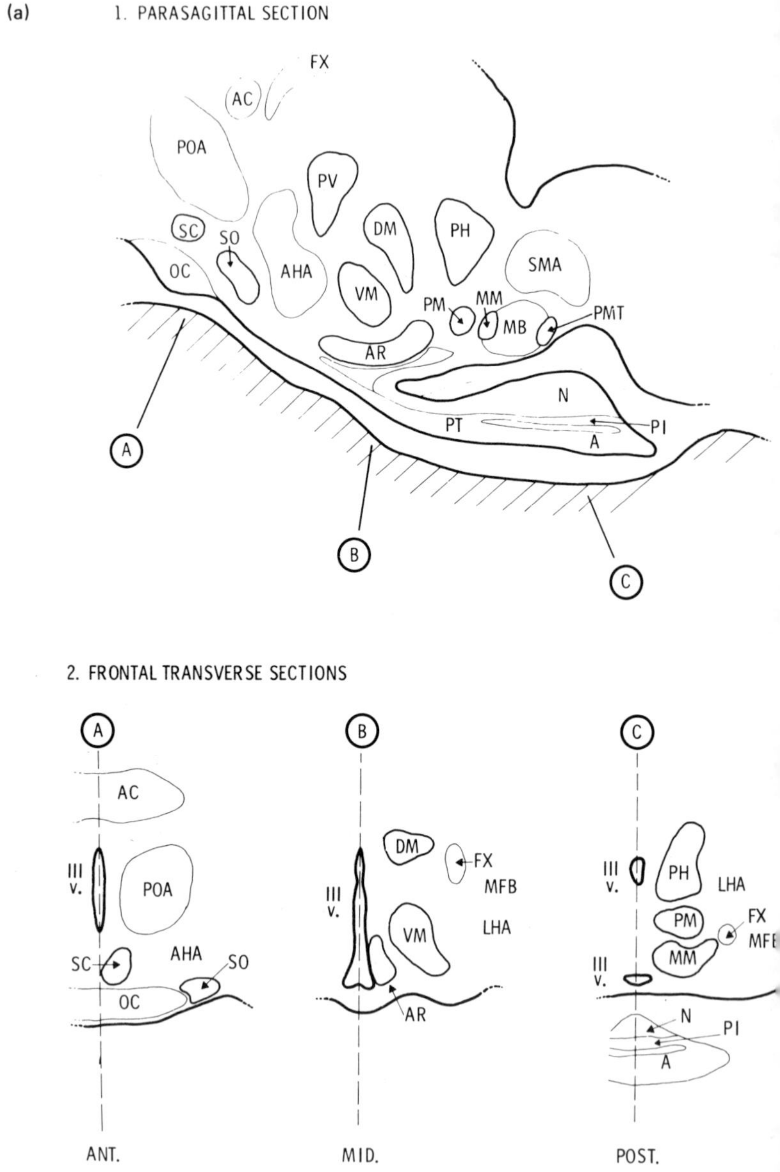
1. PARASAGITTAL SECTION
FX
AC
POA
PV
SC
SO
DM
PH
SMA
OC
AHA
VM
PM
MM
MB
PMT
AR
N
PT
PI
A
B
C
2. FRONTAL TRANSVERSE SECTIONS
A
B
C
AC
III v.
POA
SC
AHA
SO
OC
DM
FX
MFB
VM
LHA
AR
PH
LHA
PM
FX
MM
N
PI
A
ANT.
MID.
POST.

(b) 1. PARASAGITTAL SECTION

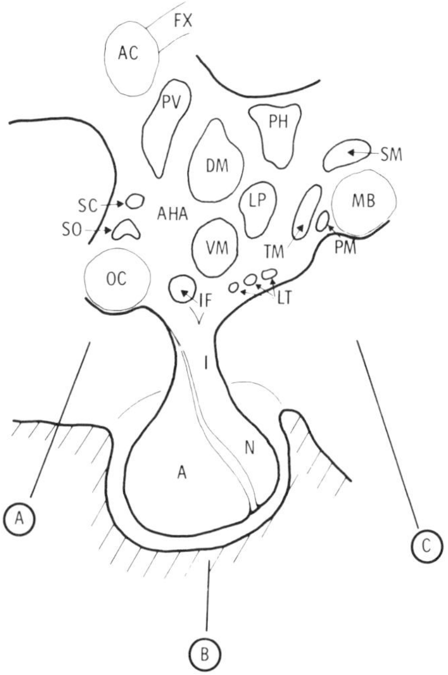

2. FRONTAL TRANSVERSE SECTIONS

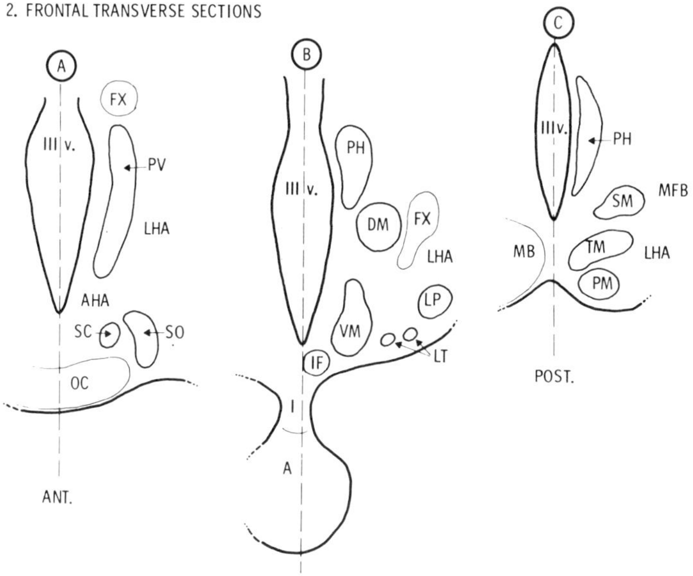

and along spinal nerves to the various visceral organs. In addition, there are regions of the hypothalamus which regulate complex patterns of behaviour including feeding, drinking, sexual activity, sleep and those associated with emotions such as anger, fear, joy, love and hate. Apart from the neural control of visceral and behavioural functions, the hypothalamus profoundly influences the endocrine system through regulation of pituitary function (see section 1:5). The hypothalamus is a highly vascular tissue (supplied by blood vessels from the circle of Willis) and a number of its functions are directly modulated by factors in blood. For example, osmoreceptors sensitive to sodium ion concentration are known to produce appropriate changes in the secretion of the antidiurectic hormone, vasopressin. There are also receptors sensitive to temperature and many other factors including peripheral hormones.

For further reading on the neuroendocrine aspects of the hypothalamus the reader is referred to [8, 12, 19, 20,] in General Book References (p. 269).

1:3 The Pituitary Gland (Hypophysis)

The pituitary gland is formed from two embryologically distinct ectodermal tissues; an evagination of the diencephalon to form the posterior lobe and an outgrowth of the buccal cavity – 'Rathke's pouch' – to form the anterior lobe. The major subdivisions of the pituitary are shown in Figure 1.3; morphologically these vary in size and relationship in different species. The infundibulum, including the median eminence, are simple structures in the rat but more complex in other species, especially in the higher primates and man. The pars intermedia derives from the inner layer of buccal ectoderm and in some species forms a cup around the neural lobe. The pars tuberalis develops as non-secretory tissue which wraps around the infundibular stem, whilst the bulk of the adenohypophysis becomes sinusoidal tissue with developing blood vessels to form the glandular pars distalis.

The blood vessels which supply the anterior pituitary were first studied in detail in the 1930s and 1940s when it was first demonstrated that the blood flowed downwards in the hypophyseal portal vessels, contrary to the previously widely held view of flow towards the brain. The demonstration of blood flow from hypothalamus to pituitary provided a framework for the concept of a neurohumoral control of anterior pituitary function, which developed in the 1950s

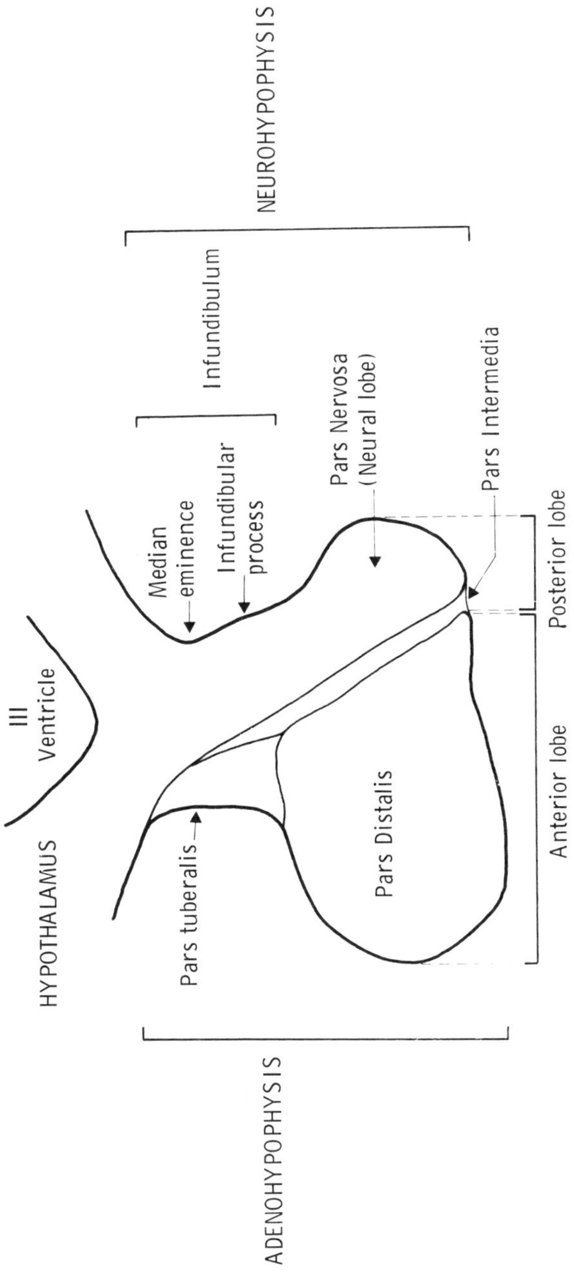

Figure 1.3: The Major Subdivisions of the Pituitary Gland.

(see section 1:5). Ironically, recent evidence indicates an upward flow in some vessels (retrograde flow) although flow towards the pituitary predominates. The blood vessels were first described by histological methods and subsequently more sophisticated techniques enabled direct observation of flow in several species. The blood supply of the pituitary is similar in most mammals in that the blood which reaches the sinusoids of the pars distalis passes via capillary loops through either long or short hypophyseal portal vessels. The superior hypophyseal artery supplies the median eminence and infundibular stem, whereas the inferior hypophyseal artery supplies the neural lobe chiefly, but also the pars distalis via the short portal vessels. The superior and inferior arteries are connected by the trabecular artery (Figure 1.4).

For further information on anterior pituitary structure and function the interested reader is referred to [7, 9, 13] in General Book References (p. 269).

Figure 1.4: The Major Blood Vessels of the Human Pituitary.

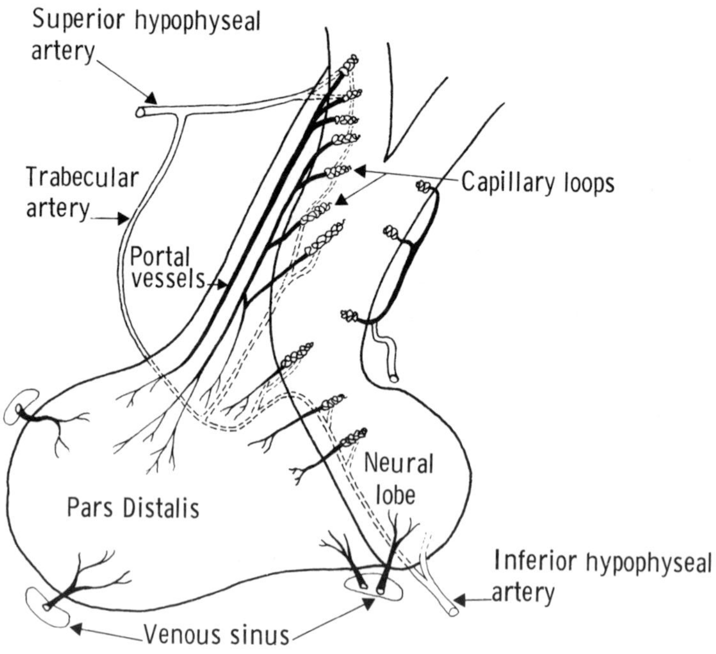

1:4 Neurosecretory Cells – Neurohormones

The predominant neurosecretory cells in the hypothalamus are those which largely terminate in the neurohypophysis (the neurohypophyseal system) and release hormones into the general circulation (see section 1.6). Other groups of neurosecretory cells in the hypothalamus mostly terminate in the median eminence (the tuberoinfundibular system) and release their products into the hypophyseal portal vessels. These hypothalamic neurosecretory cells are also peptidergic neurones since their secretory products are peptides. Recent research indicates that peptidergic neurones are widely distributed throughout the central and peripheral nervous system, but only where the peptide is secreted into blood are the neurones considered as neurosecretory.

Generally, peptidergic neurosecretory neurones are not very different from other nerve cells in the way that they synthesise, store and release a particular product, be it neurotransmitter or neurohormone. However, since the target cells of neurosecretory neurones are removed some distance from the site of release, in contrast with the postsynaptic target cells for neurotransmitters, there is a need for neurosecretory cells to manufacture large amounts of neurohormone in their cell bodies. Neurosecretory peptidergic neurones are therefore morphologically characterised by predominant protein synthetic apparatus common with other active peptide hormone-secreting cells (Figure 1.5) i.e. abundant, rough endoplasmic reticulum (RER), Golgi apparatus and membrane bound granules. This biosynthesis has been most thoroughly studied in the neurones of the neurohypophyseal system as described in detail in Chapter 3, section 3. It is assumed that such a process of synthesis and storage is common to all neurosecretory neurones, many of which are not easily studied by the techniques which have been applied to the neurohypophyseal neurones. Generally the neurosecretory peptide is synthesised in the cell body as an inactive larger precursor polypeptide by the usual process of ribosomal protein synthesis on the RER. This material is then packaged into granules in the Golgi complex together with specific peptidase enzymes, and then the granules are transported from cell body to nerve terminal down the axon by means of an active axonal transport process. During transportation and subsequent storage of the granules the active hormone or secretory product is processed from the precursor by the action of the copackaged peptidase enzymes.

Neurosecretory cells release their active hormone by mechanisms identical with those widely described for neurotransmitter release.

Figure 1.5: Principal Components of the Neurosecretory Cell.

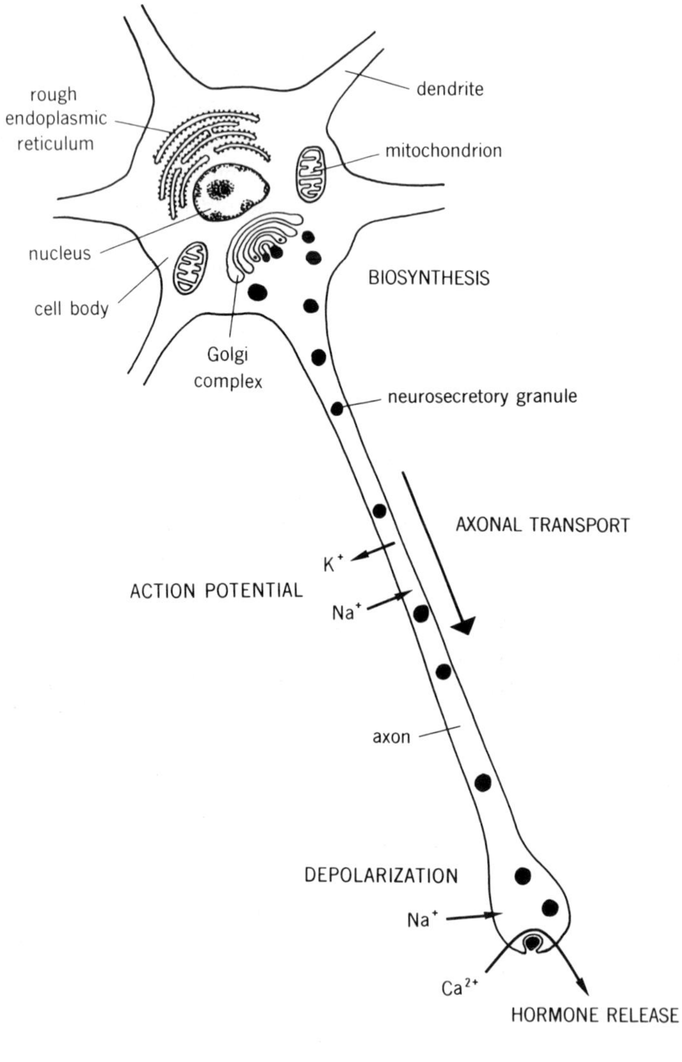

They actively maintain a membrane potential like all other neurones and the trigger for hormone release is membrane depolarisation at the nerve terminal which occurs as a result of an action potential. Depolarisation of the nerve terminal promotes the entry of calcium ions which are essential for hormone release. The release involves a mechanism in which granule membranes fuse with the nerve terminal membrane and the total granule contents are extruded – a process called exocytosis. Alternative mechanisms for release of neurohormones and neurotransmitters have been proposed but convincing evidence exists that the neurohypophyseal hormones are released by exocytosis.

For further details on neurosecretion see Maddrell & Nordman (1979) *Neurosecretion* (Blackie, Glasgow) and for information on the biotransformation and degradation of hormones and peptides see Chapter 3 section 3, Chapter 5 section 2(ii), Chapter 10 section 2 and Chapter 11 section 4, and [13].

1:5 Neurohumoral Control Mechanisms

The neural control of endocrine function is exerted in a number of ways as illustrated in Figure 1.6. This can occur through neurosecretory neurones releasing a hormone into the blood (Figure 1.6a and b) or by a neurone directly innervating a secretory cell (Figure 1.6c and d). The neurohypophyseal neurones release the two peptide hormones, oxytocin and vasopressin from the posterior pituitary into the general circulation (b). The neurones of the tuberoinfundibular system release their peptides into the hypophyseal portal blood vessels in which they are carried down to the anterior pituitary and stimulate or inhibit the release of various adenohypophyseal hormones (a). On the other hand the release of the intermediate lobe hormones (melanocyte stimulating hormone and others) is controlled by hypothalamic neurones which directly innervate the intermediate lobe secretory cells (c). In the periphery the secretion of catecholamines (adrenaline and noradrenaline) from the chromaffin cells of the adrenal medulla is controlled by acetylcholine (ACh) released from preganglionic sympathetic neurones (d). The chromaffin cells are modified postganglionic neurones and may be considered as neurosecretory cells, in that they release hormones into the general circulation.

The neurosecretory neurones of the neurohypophyseal and tuberoinfundibular systems convert neural signals into endocrine signals and can be classified as transducer cells – their inputs are electrical, their

Figure 1.6: Neural Control of Endocrine Function.

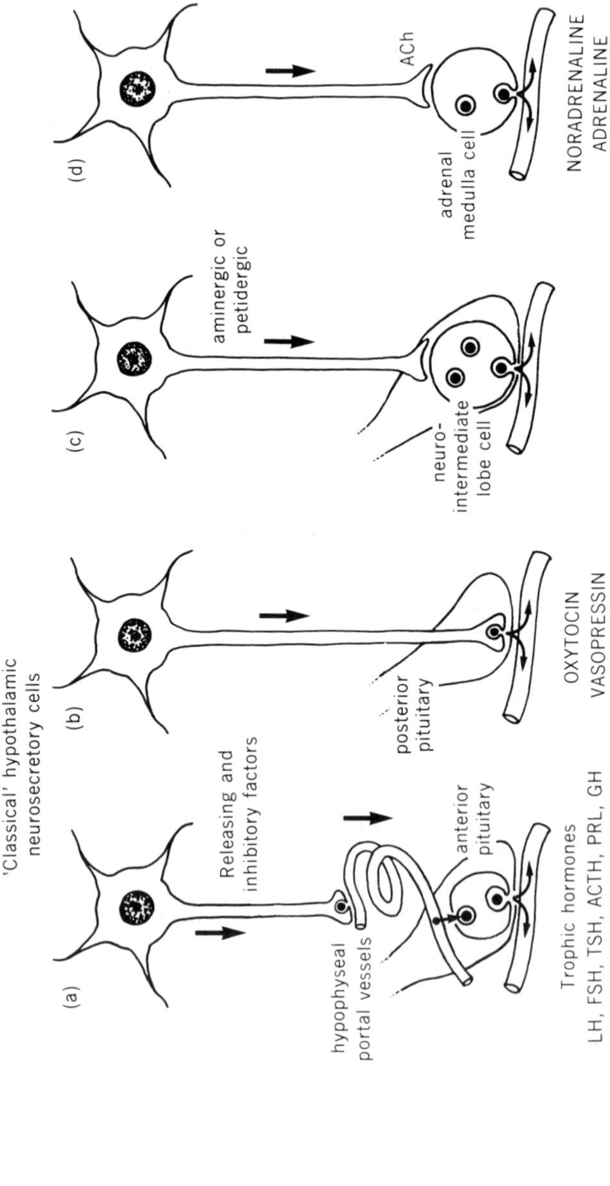

outputs hormonal. Thus their secretions will depend on any inputs that may alter their electrical excitability and thereby generate nerve impulses or action potentials which, as we have seen, results in secretion. Inputs to neuroendocrine transducer cells can be both neuronal and hormonal. Modulation of the secretion of neurohormones, occurs via neurotransmitters from other neurones which impinge on the hypothalamic neurosecretory neurones and via the feedback action of peripheral hormones (see section 1:9). We can take an example of how a variety of signals may be integrated at the level of a neuroendocrine transducer cell. Ovulation, the process of release of mature ova from the ovary, is stimulated by a surge of luteinising hormone (LH) released from the pituitary gland in response to an increased secretion of gonadotrophin releasing hormone (Gn-RH or LH-RH) from specific hypothalamic neuroendocrine cells. This surge is dependent on the feedback action of hormones secreted by the ovary but also neural signals (neurotransmitters) from the brain. In this way factors such as stress, depression, duration of daylight and tactile stimulation associated with copulation can, in particular circumstances, affect both the occurrence and timing of ovulation. Thus, integration of a broad spectrum of information operates through the final common pathway of the neuroendocrine cell and enables an extremely 'fine tuning' of a particular endocrine system. A further example of neuroendocrine transducer cells are those which occur in the pineal gland (see Chapter 9 section 2). In the following sections the main components of the hypothalamic neural and neurohumoral control mechanisms will be summarised. Further details can be found in subsequent chapters.

1:6 Control of Posterior Pituitary Hormones

As described in the previous section the chief secretory products of the posterior pituitary or neural lobe are the peptide hormones vasopressin and oxytocin (see Figure 3.1). These hormones are synthesised in two bilaterally paired nuclei in the hypothalamus – the supraoptic nuclei (SON) which produces predominantly vasopressin and the paraventricular nuclei (PVN) producing mainly oxytocin (Figure 1.2). Vasopressin and oxytocin are synthesised in the cell bodies of the SON and PVN, each with a specific carrier protein called neurophysin then transported to the neural lobe and stored prior to release. The neurophysins are single chain polypeptides of about 95 amino acids and are synthesised from part of the polypeptide precursor to the posterior pituitary hormones. The neurophysins bind in simple stoichiometric ratios with the respective neurohypophyseal hormones in the granules, and are released with the hormone although their role after release from the neural lobe is not known.

The principal action of vasopressin (hence its synonym anti-diuretic hormone – ADH) is to increase water reabsorption in the collecting ducts of the kidney although it also causes generalised vasoconstriction of blood vessels. Consistent with this action, the secretion of vasopressin is primarily controlled by variations in blood plasma osmolality. An increase in blood osmolality resulting from body dehydration or loss of water is detected by osmoreceptors in the hypothalamus which directly stimulate the secretion of vasopressin leading to water retention and hence a reduction in plasma osmolality. The secretion of vasopressin may also be increased by a loss of blood volume as occurs in haemorrhage and the retention of water and generalised vasoconstriction brought about by the actions of vasopressin help to compensate for the reduced blood pressure following haemorrhage. In the female mammal, oxytocin stimulates milk ejection and contraction of the uterus and it is released in response to two types of sensory stimuli – either suckling of the offspring or cervical stimulation. Both types of stimuli result in the transmission of afferent neural signals to the neurosecretory cells bringing about the release of oxytocin and consequent contraction of the myoepithelial cells of the mammary gland and uterine smooth muscle. These 'reflex' pathways therefore have a neural (afferent) input with a hormonal (efferent) output. Oxytocin is secreted in male mammals but its physiological actions and control mechanisms are not understood.

The neurohypophysis has also been shown to contain significant levels of a number of other neuropeptides, including the enkephalins, neurotensin, somatostatin and thyrotrophin releasing hormone (TRH), and other neurotransmitters including acetylcholine and catecholamines but the role of these agents, and whether they are also released from the posterior pituitary, is not at all clear. Note that the evolutionary aspects of the neurohypophyseal hormones, i.e. their nature and distribution in non-mammalian vertebrates, will not be covered in this book. The reader interested in the evolution of hormones and neuropeptides is referred to [1, 8] and E. J. W. Barrington (ed) (1979) *Hormones and Evolution* (Academic Press, New York).

1:7 Control of Intermediate Lobe Hormones

The main hormone secreted by the pars intermedia of the pituitary gland is α-melanocyte stimulating hormone (melanotrophin or αMSH) (see Figure 4.1). Although its role in skin colour adaptation in lower vertebrates is well known, its function in mammals is not at all clear, particularly in humans where its secretion in the adult is in doubt. The release of αMSH, both in mammals as well as lower vertebrates, is under hypothalamic inhibitory control. This control, unlike that of the pars distalis (see section 1:8) is mediated directly by neurones which terminate in the vicinity of the hormone-secreting cells and appears to involve both catecholamines and peptidergic substances (see Figure 4.2). An αMSH release-inhibiting factor (MIF) and also an αMSH releasing factor (MRF), both derived from oxytocin, have been characterised but their physiological significance remains to be demonstrated. A further intermediate lobe hormone has been described recently and called corticotrophin-like intermediate lobe peptide (CLIP) but the control of its secretion requires further study.

1:8 Control of Anterior Pituitary Hormones

The cells of the anterior pituitary (pars distalis) secrete at least eight separate hormones. These hormones may be conveniently grouped into three classes according to their chemical structures and the type of cell producing them.

(A) Corticotrophin-related peptide hormones. These hormones are secreted by chromophobic and basophilic cells of the adenohypophysis. They are all single chain linear polypeptides and include adrenocorticotrophic hormone – ACTH – (39 amino acids); melanocyte stimulating hormone – pro γ MSH – (103 amino acids); lipotrophin – βLPH – (91 amino acids). See Fig 5:2.

(B) Glycoprotein hormones. These hormones are secreted by basophilic cells and they are relatively large proteins with molecular weights in the order of 30,000. The molecules are composed of two subunits, α and β each containing a carbohydrate moiety. The smaller α unit (96 amino acids) is common to all the pituitary glycoprotein hormones, luteinising hormone (LH), follicle stimulating hormone (FSH) and thyroid stimulating hormone (TSH). The β unit (about 120 amino acids) confers biological specificity on the hormone.

(C) Growth hormone (GH) and prolactin. These two structurally related protein hormones are secreted by acidophil cells in the adenohypophysis. They are made up of about 190-200 amino acids (molecular weight approximately 20,000) and their structure is characterised by the presence of two or three disulphide bridges (forming cystine from cysteine amino acids in the peptide sequence).

In the mammal, the various secretory cells of the pars distalis are randomly dispersed throughout the sinusoidal structure. Until recently it was thought that each pituitary hormone was secreted by a distinct cell type but it is now considered likely that ACTH, βLPH and pro γ–MSH αMSH are secreted from the same cells, and the same may be true for LH and FSH.

As a result of the isolation and determination of the chemical structures of the anterior pituitary hormones, highly purified forms or, in some cases, synthetic forms of the hormones have become available. As a consequence the development of more sensitive and specific biological, immunological and cytochemical assays for the hormones has been possible (see Chapter 2 sections 3(iv) and 4(iii)). These assays have sometimes enabled accurate measurement of the basal levels of hormones in blood which has improved our understanding of the control of anterior pituitary secretions as well as the mode of action of the hormones at peripheral sites. The main peripheral actions of the anterior pituitary hormones are summarised in Table 1.2.

The secretion of anterior pituitary hormones is controlled by several specific neurohormones (also called hypophysiotrophic hormones) which act as releasing or release-inhibiting hormones. As described above they are synthesised in neurones of the tuberoinfundibular system and exert their effects on the adenohypophysis following their secretion into the hypophyseal portal capillaries. A second route by which neurohormones may reach the anterior pituitary is via the cerebrospinal fluid (CSF). Electron-microscopic studies have shown that elongated, branched cells called ependymal tanycytes, traverse the median eminence and it has been proposed that neurohormones may be secreted into the CSF and be actively transported from the third ventricle to the hypophyseal portal capillaries by the tancycytes. For further information on the transport of hormones between blood and brain via the median eminence and other so-called circumventricular organs where close contact between blood and CSF occurs see Chapter 11 section 4.

Table 1.2: Actions of the Adenohypophyseal Hormones

Hormone	Cell type	Target tissue	Rapid and metabolic actions	Slow or trophic action
Corticotrophin-related[a]				
Adrenocorticotrophic hormone (ACTH)	Corticotroph	Adrenal cortex	Synthesis and Release of corticosteroids. Increased protein phosphorylation.	Increased synthesis of steroidogenic enzymes. Overall growth of adrenal.
Lipotrophin (LPH)	Corticotroph	Adipose (fat)	Lipolysis	
Intermediate lobe hormone, Melanocyte-stimulating hormone (MSH)	Melanotroph	Skin? Other?	(Stimulates melanocytes and colour change in lower vertebrates) (see 4:2ii).	Stimulation of sebaceous and preputial (pheromone) glands in rat has been proposed.
Growth hormone-related				
Growth hormone (GH) (Somatotrophin)	Somatotroph	Liver, muscle, adipose, thymus, others? (action on bone via somatomedins)	Gluconeogenesis, lipolysis stimulation of amino acid transport. Reduction in amino acid catabolism.	Non-specific increase in synthesis of all proteins (synergistic action with other trophic hormones). Overall growth of all tissues (possible exception of developing nervous system). Lymphoid cell proliferation.

Table 1.2 contd.

Hormone	Cell type	Target tissue	Rapid and metabolic actions	Slow or trophic action
Prolactin (luteotrophin)	Lactotroph	Mammary epithelium	Stimulation of protein kinase.	Synthesis of casein and other milk protein. Overall growth, general synthesis of DNA, RNA and protein.
		Ovary		Trophic influence on corpus luteum (in some species)
Glycoprotein Hormones				
Luteinizing Hormone (LH)	Gonadotroph	Testis (Leydig cell)	Androgen synthesis and release e.g. testosterone	
		Ovary (interstitial cells, follicular cells, corpus luteum)	Oestrogen synthesis and release e.g. Oestradiol	Follicular maturation and rupture. Growth of corpus luteum.
Follicle Stimulating Hormone (FSH)	Gonadotroph	Testis (Sertoli cells)	Production of androgen binding protein, steroid conversions.	Maintenance of sperma-togenesis.
		Ovary (Granulosa cells)	Stimulates conversion of androgen into oestrogen.	Follicular growth.
Thyrotrophin (TSH) (Thyroid stimulating hormone)	Thyrotroph	Thyroid	Release of thyroid hormones (T_3, T_4). Active uptake of iodine.	Synthesis of thyro-globulin. Accelerated formation of phospholipids.

[a] There is now evidence for the secretion of a further coricotrophin-related hormone from the adenohypophysis, namely β-endorphin, but its target tissue(s) and actions have not been clarified.

Table 1.3

(A) Hypothalamic Hypophysiotrophic Hormones with known Structure and Function

Name	Structure	Function
Thyrotrophin-releasing hormone (TRH)	Pyro-Glu-His-Pro-NH_2	Releases TSH and prolactin
Gonadotrophin-releasing hormone (GnRH, luteinising hormone releasing hormone, LHRH)	Pyro-Glu-His-Trp-Ser-Tyr-Gly-Leu-Arg-Pro-Gly-NH_2	Releases LH and FSH
Somatostatin (growth hormone release inhibiting hormone, SRIH)	Ala-Gly-Cys-Lys-Asn-Phe-Phe-Trp-Lys-Thr-Phe-Thr-Ser-Cys-OH (S—S bridge between the two Cys)	Inhibits GH and TSH release

(B) Hypothalamic Hypophysiotrophic hormones with known Structure but whose Physiological Function is not Fully Defined

Name	Substance	Action
Prolactin inhibiting factor (PIF)	Catecholamine, dopamine	Inhibits prolactin secretion
Corticotrophin releasing factor (CRF)	Peptide; 41 amino acids (See 5:3)	Stimulates ACTH and β-endorphin release

(C) Hypophysiotrophic hormones with known function but unknown structure

Name	Function
Growth hormone releasing factor (GRF) (somatotrophin releasing factor)	Releases GH
Prolactin releasing factor (PRF)	Releases prolactin
Melanocyte stimulating hormone releasing factor (MSH-RF) (MRF)	Stimulates MSH release
Melanocyte stimulating hormone inhibiting factor (MSH-IF) (MIF)	Inhibits MSH release

The hypothalamic distribution of cell bodies producing the hypophysiotrophic hormones is, in general, less clearly defined than those of the neurohypophyseal system although their axonal projections predominantly terminate in the median eminence in the region of the portal capillaries. To date three of these neurohormones have been chemically identified as small peptides and their molecular structure is shown in Table 1.3 (A).

The humoral transmission of hormones from the hypothalamus to the anterior pituitary gland was not established until the early 1950s following the pioneering work of G. W. Harris in England. Nearly two decades elapsed, however, before the first hypothalamic hormone was purified and its chemical structure determined. In 1969, two research groups in North America, namely R. Guilleman and his collaborators and A. V. Schally and his collaborators, independently extracted and chemically purified thyrotrophin releasing hormone (TRH) showing it to be a small peptide containing three amino acids. This was the culmination of years of arduous work – an enormous feat involving the extraction of more than 250,000 hypothalami from sheep and pig respectively and the use of sensitive and specific bioassays to measure many fractions for hormone activity. A year later the decapeptide, gonadotrophin releasing hormone (Gn-RH or LH-RH) was subsequently characterised and in 1973 the 14 amino acid structure of somatostatin (growth hormone release-inhibiting hormone) was elucidated. Although corticotrophin releasing factor (CRF) was the first hypothalamic hormone to be described by bioassay techniques it has only very recently been isolated in pure form in sufficient amounts for clear identification (see Table 1.3B) A large polypeptide of 41 amino acids has been described but that this is the true and the only CRF still requires confirmation. At least four other hormonal factors have been recognised by physiological experiments and bioassays but their chemical structure awaits identification (see Table 1.3C). The criteria required to demonstrate that a particular substance in the hypothalamus has hypophysiotrophic function are listed in Table 1.4.

The secretion of both gonadotrophins, LH and FSH, is stimulated by gonadotrophin-releasing hormone (Gn-RH) (also termed luteinising hormone releasing hormone – LH-RH). TSH is controlled by thyrotrophin releasing hormone (TRH) and ACTH by corticotrophin releasing factor (CRF). The secretion of growth hormone (GH) and prolactin appear to be regulated by a dual control mechanism from the hypothalamus in that there is evidence for both releasing and release-inhibiting factors. Somatostatin is the hormone which inhibits

Table 1.4: Criteria[a] Required to Demonstrate that a Particular Substance Acts as a Releasing or Release-inhibiting Hormone

1. The substance releases or inhibits the release of the pituitary hormone preferentially from the anterior pituitary both
   (a) in whole animal preparations (*in vivo*) and
   (b) in isolated tissue preparations (*in vitro*)
2. The substance is present in the hypothalamus and hypophyseal portal blood in greater concentrations than in peripheral (systemic) blood.
3. Its concentration in the hypothalamus and portal blood varies with physiological influences consistent with the action of such influences on the secretion of the pituitary hormone.
4. Its action on its pituitary hormone secretion is inhibited by the target hormones of that hormone (e.g. corticosteroid inhibition of ACTH secretion).
5. The electrical activity of the hypothalamic cells which produce the substance are modulated by both pituitary and target hormones.
6. Mechanisms exist in the hypothalamus to synthesise the substance and mechanisms to inactivate (or degrade) the substance occur at pituitary and other sites.

[a] Substances listed in Table 1.2 satisfy most, if not all, these criteria.

Figure 1.7: Control of Anterior Pituitary Secretion.

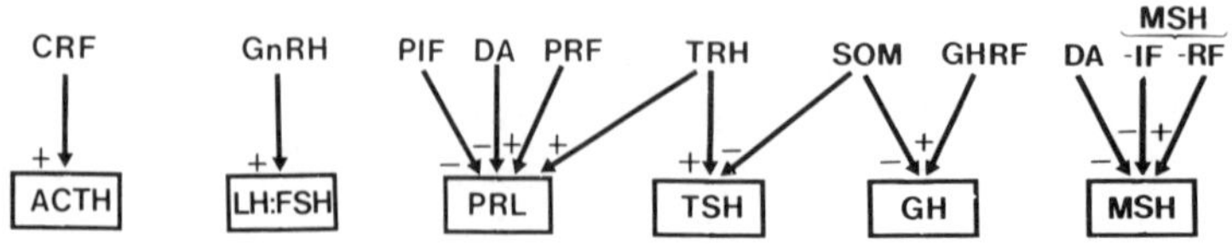

Key

Pituitary Hormones

ACTH	Adrenocorticotrophin		
LH	Luteinising hormone	DA	Dopamine
FSH	Follicle stimulating hormone	PIF	Prolactin release inhibitory factor
PRL	Prolactin	PRF	Prolactin releasing factor
TSH	Thyroid stimulating hormone	TRH	Thyrotrophin releasing hormone
GH	Growth hormone	SOM	Somatostatin
MSH	Melanocyte stimulating hormone	GHRF	Growth hormone releasing factor

Hypothalamic Hormones/Factors

CRF	Corticotrophin releasing factor	MSH-IF	MSH release inhibiting factor
Gn-RH	Gonadotrophin releasing hormone	MSH-RF	MSH releasing factor

the release of GH and, although the predominant hypothalamic influence on GH release is stimulatory, a GH releasing factor has not yet been identified. In contrast to the hypothalamic control of GH secretion, the hypothalamic control of prolactin release in basal conditions is predominantly inhibitory in nature. One of the major prolactin release-inhibiting hormones is the catecholamine, dopamine although other peptide factors have been shown to inhibit or stimulate prolactin secretion. As shown in Figure 1.7 there is 'cross talk' between the hypothalamic hormones and their respective adenohypophyseal hormones. Experiments have shown TRH can stimulate the secretion of prolactin by a direct action on the pituitary and somatostatin can inhibit the stimulatory effect of TRH on TSH secretion. Whether such actions reflect true physiological control mechanisms, however, remains to be established.

The 'neuroendocrine' control mechanisms described in this section are outlined in more detail in later chapters. Also the interested reader is referred to a number of reviews in [3, 6, 7, 18].

It is generally evident that the hypothalamic hormones are small peptides which cause (i) acute release (or inhibition) of the anterior pituitary hormone, (ii) an increase in the synthesis of the pituitary

Table 1.5: Several Hypothalamic Peptides with Known Structure and Actions but whose Hypothalamic Function is not Fully Understood (see 10:2)

Name	Structure
Substance P	Arg-Pro-Lys-Pro-Gln-Gln-Phe-Phe-Gly-Leu-Met-NH_2
Neurotensin	Pyro-Glu-Leu-Tyr-Glu-Asn-Lys-Pro-Arg-Arg-Pro-Tyr-Ile-Leu-OH
Angiotensin II	Asp-Arg-Val-Tyr-Ile-His-Pro-Phe
Leu-enkephalin	Tyr-Gly-Gly-Phe-Leu-OH
Met-enkephalin	Tyr-Gly-Gly-Phe-Met-OH
β-Endorphin[a]	Tyr-Gly-Gly-Phe-Met-Thr-Ser-Glu-Lys-Ser-Gln-Thr-Pro-Leu-Val-Thr-Leu-Phe-Lys-Asn-Ala-Ile-Val-Lys-Asn-Ala-His-Lys-Lys-Gly-Gln-OH
γ-Endorphin[a]	Tyr-Gly-Gly-Phe-Met-Thr-Ser-Glu-Lys-Ser-Gln-Thr-Pro-Leu-Val-Thr-Leu-OH
Vasoactive inhibitory polypeptide (VIP)	His-Ser-Asp-Ala-Val-Phe-Thr-Asp-Asn-Tyr-Thr-Arg-Leu-Arg-Lys-Gln-Met-Ala-Val-Lys-Lys-Tyr-Leu-Asn-Ser-Ile-Leu-Asn-NH_2
Cholecystokinin-8 (CCK-8)	Asp-Tyr-Met-Gly-Trp-Met-Asp-Phe-NH_2 (S^1O_3H)

[a] These are also synthesised and secreted by the anterior pituitary (see Chapter 5 section 2) and the intermediate lobe of the pituitary (see Chapter 4 section 5).

hormone and (iii) in the case of Gn-RH at least, there is evidence for a priming effect, in that a second pulse of Gn-RH causes a greatly enhanced secretion of LH. With the exceptions of GH and prolactin, all the anterior pituitary hormones regulate the function of other endocrine glands, and their secretion rates are controlled by the feedback action of the hormones, release by their target endocrine gland (see section 1:9) coupled with the integrated information from various sensory neural inputs. The integration of hormonal feedback and neural information occurs chiefly at the level of the hypothalamus (and higher brain regions) and regulates the rates of release of the hypothalamic hormones. However, there is also evidence to show that the feedback of some circulating hormones can act at the level of the pituitary gland to affect its responsiveness to the hypothalamic hormones.

In addition to the established hypophysiotrophic hormones, the hypothalamus contains numerous other biologically active peptides with known structure and action on the anterior pituitary, but their possible physiological function as hypophysiotrophic hormones or regulators of other hormones is poorly understood. These peptides are shown in Table 1.5 and many are widely distributed throughout the brain, alimentary tract and other tissues (see Chapter 10 sections 1 and 2).

A number of biogenic amine neurotransmitters present in the hypothalamus converge on the median eminence and directly or indirectly influence the secretion of the anterior pituitary hormones (Table 1.6). A major aminergic pathway which terminates in the median eminence is the intrahypothalamic dopamine pathway, the tuberoinfundibular tract whose cell bodies are situated in the arcuate and ventromedial nuclei (see Figure 1.2). The neurones have both a neurosecretory function in that they can release dopamine into the hypophyseal portal capillaries and may also modulate the release of other hypothalamic hormones, as demonstrated by pharmacological studies and the existence of axo-axonic connections with peptidergic nerve terminals in the median eminence. In addition, the hypothalamus is innervated by nor-adrenergic nerve endings of the medial forebrain bundle, which is a fairly diffuse pathway arising from neurone cell bodies located in the pons and medulla of the brainstem. Other components of this ascending pathway of neurones include dopamine, adrenaline, 5-hydroxytryptamine (5-HT) and acetylcholine (ACh) containing fibres, some of which terminate in the hypothalamic area.

Figure 1.8 indicates at least four possible mechanisms by which

Table 1.6: Possible Role of Brain Neurotransmitters in Regulation of Anterior Pituitary Secretion

Neurotransmitter	Corticotrophin (ACTH)	Growth Hormone (GH)	Prolactin	Gonadotrophins (LH/FSH)	Thyrotrophin (TSH)
Dopamine	–	+–	––	+–	+–
Noradrenaline	––	++	+–	+	+?
Serotonin	+–	++	++	–+	+–
Acetylcholine	++	0	–	+	?
Histamine	+–	?	+–	+–	+–
γ-Aminobutyric acid (GABA)	–	+	–	+	–

+ = stimulation; – = inhibition; 0 = no apparent regulation; ? = unknown regulation.
The net responses shown here do not reflect precisely the intricate mechanisms which operate but are based on numerous and often conflicting data in the literature and should not be considered conclusive. Also differences between mammalian species are often observed.

β-Lipotrophin is not included since at the present time neurotransmitter regulation of the secretion of this hormone is uncertain.

Figure 1.8: Possible Mechanisms by which Neurotransmitters Modulate Anterior Pituitary Function. (a) Action via one or several neurones in a neural pathway converging on the neurosecretory neurone. (b) Direct synapse on the soma or dendrite of the neurosecretory neurone. (c) Axo-axonic synapse on the nerve endings of the neurosecretory neurone. (d) Neurotransmitter neurones secrete directly into portal vessels and neurotransmitter acts directly at pituitary level.

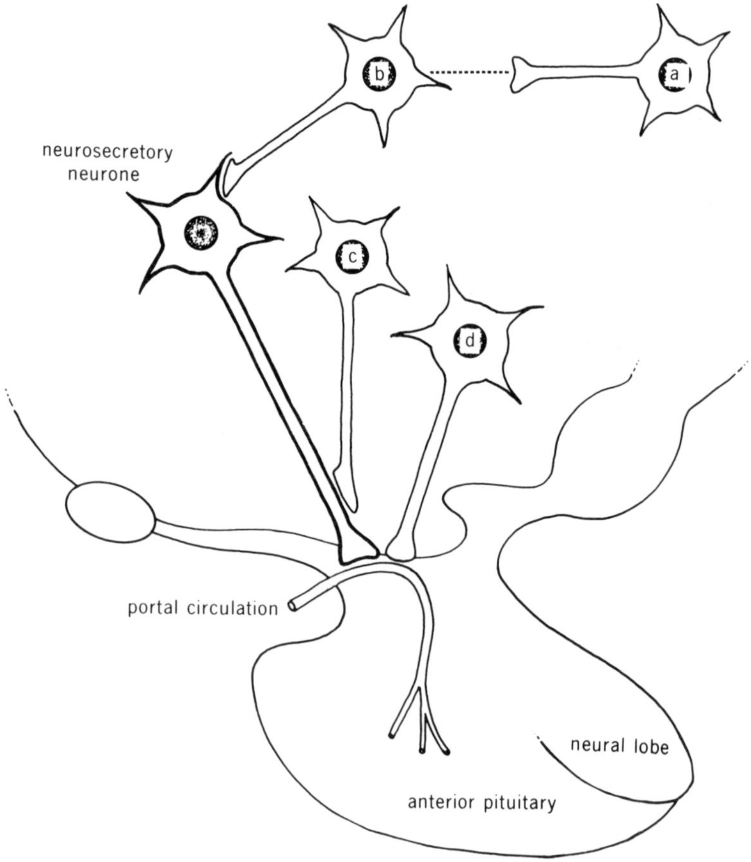

neurotransmitter modulation of pituitary hormone secretion might operate and physiological and pharmacological evidence for all these mechanisms exists. Much of the evidence for neurotransmitter regulation of anterior pituitary secretion will be presented in later chapters, but the net responses are summarised in Table 1.6. A further possible regulatory mechanism, not represented in Figure 1.8, which needs some consideration, is modulation of hormone secretion by a neurotransmitter which co-exists within the neurosecretory peptidergic neurone. Recent research has shown that certain neurones releasing neurotransmitters such as dopamine and 5-hydroxytryptamine, also contain and apparently release specific neuropeptides. Although examples of co-existence are widespread throughout the mammalian nervous system, it has not been described in the hypothalamus. This is perhaps somewhat surprising since this brain region contains the highest concentration of neurones containing amines or neuropeptides. Evidence is emerging that other possible neurotransmitters present in the hypothalamus may also have a part to play in the control of anterior pituitary function. These include the amino acid transmitters, in particular γ-aminobutyric acid (GABA), other amines such as adrenaline (which may be responsible for some of the roles described for noradrenaline) and other neuropeptides with, as yet, unknown hypophysiotrophic actions. For further reading on the neuroendocrine roles of neurotransmitters see [11, 15, 20] and for a recent review of the general properties of neurotransmitters see [10].

1:9 Feedback Mechanisms in the Control of Pituitary Hormone Secretions

In the previous section the importance of feedback mechanisms in controlling pituitary hormone secretions was described and, in order to understand individual control systems in neuroendocrinology the basic principles of hormone feedback will now be briefly considered. The function of a hormone is to regulate the metabolic activity of its specific target cells and to do this an endocrine gland requires constant and rapid information on the state of the system it is regulating. The concept of regulatory mechanisms initially arose from the work of Claude Bernard who, by 1859, had appreciated the importance of maintaining the constancy of the 'milieu interieur'. The development of control system theory in the 1940s provided a further stimulus and framework in which to describe physiological and biochemical regulatory mechanisms although, due to the non-linear nature of biological control systems, the quantitative application of this theory is limited. There are two major categories of control systems – 'open loop' in which the input is not affected by the output and 'closed loop' systems in which part of the output is fed back and participates as an input to the system. Most physiological homeostatic mechanisms are closed loop systems operating via the principle of negative feedback.

Outlined in Figure 1.9(a) is the basic form of negative feedback. It shows that the value of a regulated variable, Y, (for the present purpose, the circulating levels of a hormone secreted by a target gland) is sensed and then compared with a set point or command signal. Any difference between the feedback signal and the set point will result in an error signal so that the value of Y can be returned to the required level. Thus any increase in the controlled variable may lead to a decrease, and vice versa, thereby maintaining the level of a hormone within a narrow range. However, the simple feedback systems described below do not give a complete qualitative picture of hormonal control as an endocrine gland receives signals from a variety of inputs and the output will be determined by the integration of all these afferent signals.

The simplest negative feedback system which operates to control the secretion of pituitary hormones is a direct feedback which relates the rate of release of a pituitary hormone to the blood concentration of its target gland hormone. For example TSH, secreted by the pituitary gland, stimulates the synthesis and release of thyroid hormones from the thyroid gland. The level of 'free' circulating thyroid hormone is then sensed by the pituitary gland and the hormone can thereby regulate its own release rate by negative feedback action on the rate of TSH secretion (see Figure 1.9(b)). The set point or command signal may be considered as the release of TRH from the hypothalamic neurosecretory cell.

There are of course feedback mechanisms which operate at the level of the central nervous system, particularly the hypothalamus. Thus a hormone secreted by a target gland can have both a direct feedback on the adenohypophyseal hormone and an indirect feedback on the release of the hypothalamic hormone. This involvement of the hypothalamus is particularly important in the control of hormones secreted by the adrenal cortex and the gonads and an example of the feedback mechanisms which control the function of the adrenal cortex is illustrated in Figure 1.9(c). There is some evidence for the existence of a short feedback loop in the control of pituitary function whereby a pituitary hormone has a negative feedback action on the secretion of a hypothalamic releasing or release-inhibiting factor. Prolactin is one hormone which is thought to regulate its own secretion by a short feedback mechanism and the principle is illustrated in Figure 1.9(d). Although there has been some doubt with respect to the importance of such a short feedback mechanism, the more recent evidence of

Figure 1.9: Feedback Mechanisms in the Control of Adenohypophyseal Hormone Secretions. (a) Basic form of a negative feedback system. (b) Control of thyroid stimulating hormone (TSH) secretion by feedback effects of thyroid hormones on the pituitary. (c) Control of adrenocorticotrophic hormone (ACTH) by feedback action of cortisol on both the hypothalamus and pituitary gland. (d) Short loop autoregulatory feedback of prolactin: TRH – thyrotrophin releasing hormone; CRF – corticotrophic releasing factor; PIF – prolactin inhibiting factor; DA – dopamine.

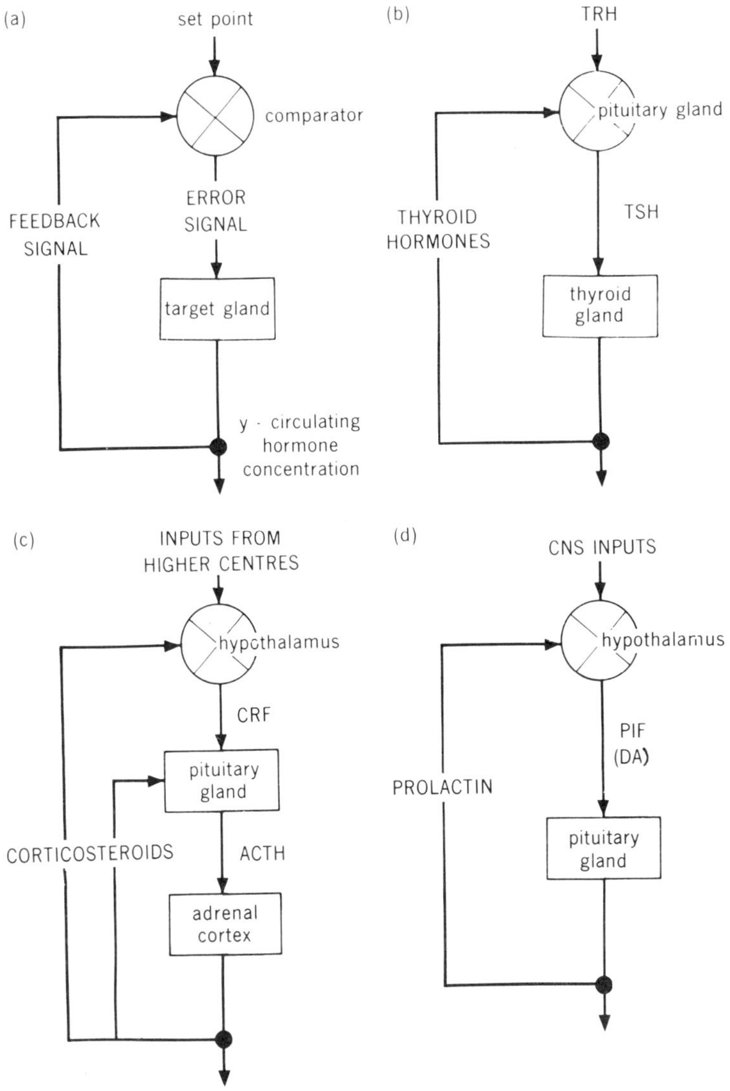

retrograde blood flow in certain hypophyseal portal vessels has added more credence to this idea. However, a short loop feedback could just as well be mediated by the general circulation.

There is at least one example in the endocrine system of a positive feedback mechanism. These are extremely rare in biological systems because basically they are unstable and lead to an 'explosive' situation when no other factors are involved. The principle of positive feedback is that an increase in a hormone concentration leads to an enhancement of its effect by stimulating further release. This situation occurs prior to ovulation when oestrogens are secreted in increasing amounts by the developing ovarian follicle(s). Normally these steroids have a negative feedback action on pituitary gonadotrophin secretion but, when they reach a critical plasma concentration they stimulate the release of gonadotrophins from the pituitary. This results in a surge of LH and FSH secretion which act on the ovary to cause follicular rupture and the release of the ovum. The reason why this positive feedback does not become uncontrollable is that by some unknown mechanism the feedback effect of oestrogen reverts to a negative feedback action thereby stabilizing the system.

The close integration between nervous and endocrine systems is emphasised by the various feedback mechanisms which operate to control pituitary secretions. The feedback loops concerned with the control of adenohypophyseal hormones involve hormones secreted by their target glands whereas those regulating posterior pituitary hormones involve neural feedback control. In both cases, the important feature of the regulation of hypophyseal hormone release is the ability of other brain regions to override or finely adjust these feedback mechanisms which operate principally at the level of the hypothalamus. For further reading on steroid hormone feedback mechanisms see [2, 14]. Also see section 1:11.

1:10 Mechanism of Action of Peptide Hormones

The receptors for peptide and protein hormones are located on the plasma or cell membrane of target cells. A variety of protein hormones, including LH, FSH, ACTH, TSH, TRH, Gn-RH and vasopressin produce their biochemical effects by stimulating an increased intracellular formation of adenosine 3′,5′ -monophosphate (cyclic AMP). Binding of these hormones to their receptors causes the activation of the enzyme, adenylate cyclase, which is located on the inner surface of the cell

Figure 1.10: Mechanism of Peptide Hormone Action. Mechanism by which cell membrane receptors may mediate hormone effects by changes in Ca^{2+} mobilisation and cyclic AMP (cAMP). AC-adenylate cyclase; H-hormone; REC-receptor; GTP-guanosine triphosphate; R-GTPase-regulatory unit containing GTP binding site and GTPase; GDP-guanosine diphosphate; P-phosphate; PK-Reg-protein kinase regulatory unit; PK-Cat-protein kinase catalytic unit; PDE-phosphodiesterase; CM-calmodulin; ATP-adenosine triphosphate.

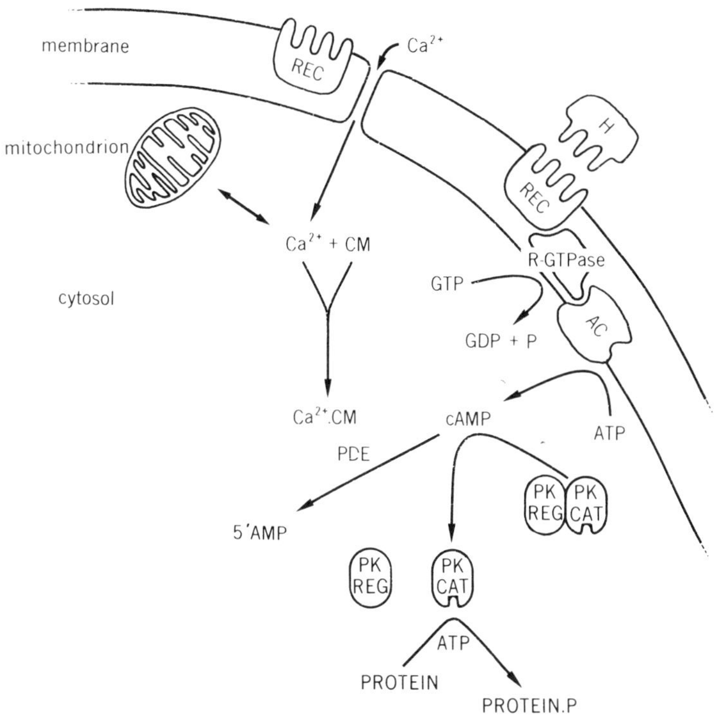

membrane (Figure 1.10). Adenylate cyclase converts adenosine triphosphate (ATP) into cyclic AMP and in turn cyclic AMP can activate a protein kinase by combining with a regulatory subunit of the kinase and promoting dissociation of the catalytic subunit. The 'free' enzymatically active subunit has the ability to transfer the phosphate from ATP to a substrate and by this process of phosphorylation can activate proteins which are important in the control of cell function. More recent evidence has suggested that the coupling of hormone to its receptor and stimulation of adenylate cyclase appears to involve a guanosine triphosphate (GTP) binding component which contains a

GTPase and breaks down GTP to guanosine diphosphate (GDP); intracellular GTP is thus thought to regulate the hormone receptor-cyclase coupling.

Hormone binding to cells is not only associated with raised levels of cAMP but also with an increased intracellular concentration of free calcium ions (Ca^{2+}) which, like cyclic AMP, are also thought to be an intracellular regulator of biochemical responses to hormones. The source of this calcium may be from passive diffusion across the cell membrane, or release from intracellular stores, or by inhibition of mitochondrial uptake. In the case of cAMP it seems likely that the regulatory subunit of a protein kinase is the only intracellular constituent with which the nucleotide has to interact for it to bring about diverse cellular responses. No single target enzyme activity of this type has been found for cytosol Ca^{2+} ions although it appears that a Ca^{2+} binding protein, calmodulin, can activate a substantial variety of enzymes when complexed with this divalent ion. Among those described to date are cyclic nucleotide phosphodiesterase, brain adenylate cyclase, and a protein kinase of plasma membranes. An attempt to interrelate hormone receptors linked with cAMP and Ca^{2+} mobilisation is schematically illustrated in Figure 1.10 and may be used as a generalised, if not highly simplified, model of many hormone receptor interactions which occur at the level of the cell membrane and involve cAMP and Ca^{2+}. However, it should be remembered that cyclic guanosine monophosphate (cGMP) has also been implicated in mediating the effects of peptide hormone–receptor interactions but its precise involvement in regulating cellular actions is as yet poorly defined.

Prostaglandins, which are derivatives of arachidonic acid and produced in many tissues (see Table 1.1) have also been implicated in mediating the action of pituitary hormones since prostaglandins were shown to mimic the effects of several pituitary hormones on their target cells. It has been suggested that prostaglandins may either couple hormone receptor activation with adenylate cyclase or that prostaglandins may act independently to stimulate cyclic AMP formation. The question as to whether or not prostaglandins are essential intermediates in the action of pituitary trophic hormones has roused considerable controversy although detailed studies on the action of LH on gonadal steroidogenesis support the concept that prostaglandins act independently of the LH receptor activation. For further information on prostaglandins see review [17].

In summary, the cyclic AMP is considered to act as a 'second messenger', mediating the effects of peptide hormone–receptor interactions to the intracellular metabolism of target cells. The specificity of the hormone action is determined by the subsequent activation of the specific proteins which regulate appropriate cell functions. Although this may be considered as a generalised mechanism of many protein and peptide hormones notable exceptions which do not appear to utilise cAMP are insulin secreted by the endocrine pancreas and oxytocin. Other hormones such as growth hormone and prolactin appear to stimulate cAMP in some but not all target tissues. For further details on the mechanisms of action of the hypothalamic hormones see [12].

It is now clear that certain peptide hormones can also be internalised in their target cells by endocytosis, the reverse of exocytosis, and this process is mediated by their specific receptors. Thus membrane receptors may not only be responsible for initiating target-cell responses via stimulation of cyclic AMP formation but also for directing the entry of hormones into cells. Binding studies of radioactively labelled peptide hormones have demonstrated that internalised hormones become associated with lysosomes or Golgi structures where the hormone receptor complex is either degraded or reprocessed. The internalisation of hormone–receptor complexes has led to speculation as to the significance of this process and it has been suggested that it may play a critical part in mediating the sensitivity of target cells to subsequent hormone action. Internalisation may reduce the number of target cell receptors or alter the rate of removal and degradation of a particular hormone. Furthermore the discovery of intracellular peptide hormone binding sites has suggested that certain slow, trophic actions of hormones may be mediated by internal receptors and recent evidence is now accumulating to suggest that some peptide hormones become slowly associated with nuclei of target cells. These findings have opened up exciting new areas of research and the implication that peptide hormones may induce their trophic effect intracellularly is analogous to the action of steroid hormones which are well known to bind to intracellular receptors and to be translocated to the nucleus. The interested reader should see [9] for further information.

Some peptides which demonstrate internalisation include Gn-RH, prolactin, the placental glycoprotein, chorionic gonadotrophin, and the pancreatic hormone, insulin. This mechanism may be more widespread since it is known that many hormones can regulate their own receptor numbers (e.g. by altering the conformation of unoccupied receptors). Thus, overstimulation of target cells by a hormone leads to an attenuation of the response by reduction in receptor density. Such an effect is termed down-regulation of receptors. Numerous pharmacological studies of neurotransmitters in brain have also shown a reciprocal effect. When neurotransmitter is removed by lesion or denervation techniques (see Chapter 2 section 3(ii)), the target cells become supersensitive to exogenous neurotransmitter. This effect has been termed up-regulation of receptors and there is evidence for a similar mechanism occurring in certain endocrine systems although such systems are frequently adjusted by further compensatory mechanisms. For further reading see [5, 9, 19]

1:11 Mechanism of Action of Steroid Hormones

We have heard that the main peripheral hormones secreted by the adrenal cortex and the reproductive glands (gonads) are steroid hormones. These hormones have important roles in various tissues to regulate a variety of physiological functions but primary amongst these roles is the feedback control of central neuroendocrine systems and for this purpose receptor-mediated mechanisms are present in pituitary, hypothalamus and other brain regions (see Chapter 5 sections 5 and 6 and Chapter 8 sections 4 and 5). Similar mechanisms in brain also mediate the behavioural responses to specific steroid hormones (see Chapter 11 sections 2 and 3). Whilst peripheral steroid-receptor mediated actions have been studied the most thoroughly (see Chapter 8 section 5), evidence suggests that similar processes operate in all steroid target tissues.

Steroids are small planar molecules with lipophilic properties (see Table 1.1 and Figure 5.1). This means that steroid hormones readily traverse cell membranes and enter the target cells. The use of radioautographic (see Chapter 2 section 4v) and subcellular fractionation techniques (see Chapter 2 section 5) has led to the general two step model of steroid hormone–receptor interaction outlined in Figure 1.11. Following entry of the steroid into the cell it combines rapidly with specific cytoplasmic receptors which have high affinity (dissociation constants 10^{-8}–10^{-10} M) but low capacity. The cytosol receptor–hormone complex is then translocated to the nucleus where it combines with a nuclear acceptor. Before this second step occurs, however, a temperature-dependent activational process is required in order that the steroid–receptor complex acquires a propensity for binding to the nucleus. The nature of the nuclear acceptor sites is a debated issue but evidence in favour of both DNA and non-histone

Figure 1.11: Mechanism of Steroid Hormone Action.

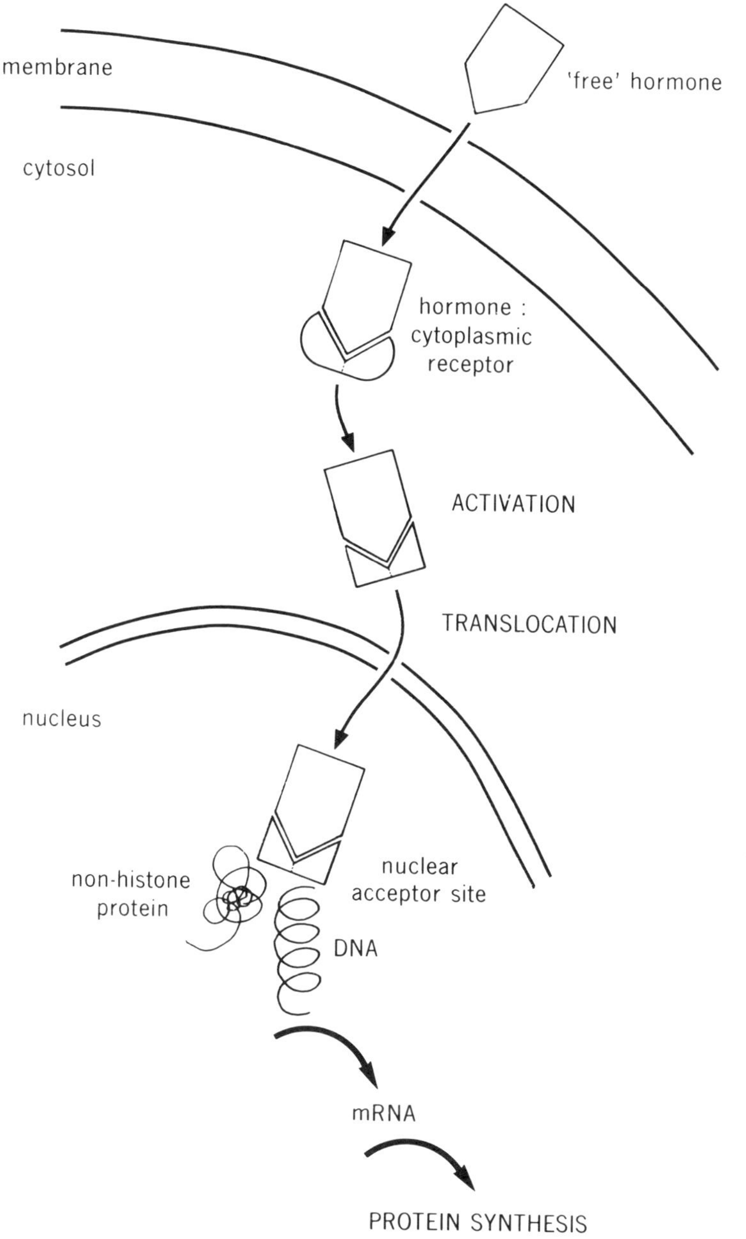

proteins has been presented. As a result of this mechanism the steroid may induce the synthesis of specific proteins in the target cells. Although this mechanism of steroid action is thought to operate in all target tissues and mediates their long-term effects, there is evidence that steroid hormones in brain, and possibly other tissues, also demonstrate acute actions which do not involve protein synthesis. For further reading on steroid hormone action see [2, 14] and V. H. T. James & J. R. Pasqualini (eds) (1980) *Hormonal Steroids* (Pergamon Press).

1:12 Neuroendocrine Integration

In the preceding sections the concept of the neurosecretory cell as the ultimate link between nervous and endocrine systems has been discussed. However, during the past decade there have been considerable developments in the field of neuroendocrinology which have led to a re-evaluation of the classical distinction between nerve cells and endocrine cells. It is not appropriate to discuss all the recent developments, some of which are covered in more detail in subsequent chapters, but here we shall highlight some of the evidence which is beginning to unify concepts held by neurochemists, neurophysiologists, pharmacologists and endocrinologists and in some instances pose a possible requirement for new definitions and terminology.

Following the characterisation of TRH, GnRH and somatostatin in the hypothalamus and the development of sensitive and specific radioimmunoassays for these peptides, it was shown that the peptides were present not only in hypothalamic tissue but also in other parts of the brain and spinal cord. Furthermore, significant concentrations of somatostatin and also TRH have been found in the gastrointestinal tract, pancreas, retina and other tissues. It is now well recognised that these peptides have many specific biological actions other than controlling anterior pituitary secretion and such actions range from behavioural effects to the modulation of peripheral hormone secretion.

Apart from the wide distribution of the so-called hypothalamic hormones, a host of other biologically active peptides have been shown to exist within both the central nervous system (CNS), the peripheral nervous system, the pituitary gland and, in many cases, also in the gastrointestinal tract. To date there are some 30 biologically active peptides present in the brain and gut and these include the neurohypophyseal hormones, some of the anterior pituitary hormones, endorphins, substance P, neurotensin, angiotensin II, vasoactive intestinal polypeptide (VIP) cholecystokinin (CCK), bradykinin, gastrin, bombesin and pancreatic polypeptide – and the list increases year by year. As implicated in the previous section experiments have shown that a number of these peptides (see Table 1.3) are present in high concentrations in the hypothalamus and may have important neuroendocrine functions.

The existence of peptidergic neural pathways appears to occur early in phylogenetic history however, and it would seem that certain functions such as the hypothalamic control of anterior pituitary secretion, have co-opted a primitive molecule for its purpose. TRH, for instance, is present in the brains of the earliest chordates e.g. Amphioxus, although this species has no TSH or pituitary gland. Also in the earliest land vertebrates, the amphibia, TRH is abundant in CNS tissue including the hypothalamus but does not appear to regulate pituitary–thyroid activity in many of the amphibia species studies. Thus, in evolutionary terms, the hypothalamic peptides may have originated from peptidergic neurones which were unrelated to the control of pituitary secretion.

There is a growing body of evidence which suggests that some of these peptides in the CNS are released and act as neurotransmitters. This has been extensively investigated with respect to substance P and the enkephalins as transmitters in the spinal cord involved with the regulation of pain transmission. On the other hand, neuropeptides may simply function as modulators of neurotransmission but only future research, which promises many exciting developments, will unravel the true nature and roles of these peptides in the CNS.

The evidence that similar peptides are present in the gastro-intestinal tract has provided a parallel field of study of these peptides. An explanation for the homology of peptides in gut and brain appeared in 1968, when A. G. E. Pearse proposed a visionary concept of APUD cells (APUD is an acronym standing for amine content and/or amine precursor, uptake and decarboxylation). He observed that neurones and certain endocrine cells producing polypeptide hormones shared similar ultrastructure and cytochemical features and suggested APUD could summarise common features of these cells. Furthermore, he suggested that embryologically all peptide-secreting cells are derived from specialised neural ectoderm tissue which during development, forms the neurones, the neurosecretory cells of the hypothalamus, and the peptide secreting cells of pituitary and gut. Evidence in favour of the APUD concept leads to the premise that all secretory cells are basically similar and that the classically defined nerve cells and endocrine cells, which at face value appear to function very differently, simply represent extremes of a spectrum which includes neurosecretory and paracrine cells (see Figure 1.1). Although this concept may require modification in light of recent evidence (see [15]) much information on the function of peptides in the gut, be it endocrine, paracrine or neurotransmitter may further our understanding of the neuroendocrine role of these peptides. However, detailed discussion of the gastrointestinal tract will not be considered in this book (see S. R. Bloom & J. M. Polak (1981) *Gut Hormones* (Churchill Livingstone, Edinburgh). The recent research developments, described here and in more detail in subsequent chapters, have placed neuroendocrinology and our understanding of neuroendocrine integration in the mammal into a far wider context than was previously imagined.

References

1. Acher, R. (1981) Evolution of neuropeptides. *Trends in Neurosciences*, **4**: 225-9.
2. Fink, G. (1979) Feedback action of target hormones on hypothalamus and pituitary with special reference to gonadal steroids. *Annual Review of Physiology*, **41**: 571-80.
3. Flerko, B. (1980) The hypophysial portal circulation today. *Neuroendocrinology*, **31**: 56-63.
4. Fujita, T. & Kobayashi, O. (1979) Current views on the paraneurone concept. *Trends in Neurosciences*, **2**: 27-30.
5. Greengard, P. (1978) *Cyclic nucleotides, phosphorylated proteins and neuronal function.* (Raven Press, New York).
6. Guillemin, R. (1978) Biochemical and physiological correlates of hypothalamic peptides: the new endocrinology of the neuron, in S. Reichlin, R. J. Baldessarini and J. B. Martin (eds), *The Hypothalamus* (Raven Press, New York), pp. 155-94.
7. Harris, G. W. (1972) Humours and Hormones. *Journal of Endocrinology*, **53**: ii-xxiii.
8. Jackson, I. M. D. (1980) Distribution and Evolutionary Significance of the Hypophysiotropic Hormones of the Hypothalamus. *Frontier Hormone Research*, **6**: 35-9.
9. King, A. C. & Cuatrecasas, P. (1981) Peptide hormone-induced receptor mobility, aggregation and internalization. *The New England Journal of Medicine*, **305**: 77-88.
10. Kruk, Z. L. & Pycock, C. J. (1979) *Neurotransmitters and Drugs*. Croom Helm (London).
11. Krulich, L. (1979) Central neurotransmitters and the secretion of prolactin, GH, LH and TSH. *Annual Review of Physiology*, **41**: 503-616.
12. Labrie, F., Borgeat, P., Drouin, J., Beaulieu, M., Lagace, L., Ferland, L. & Raymond, V. (1979) Mechanism of action of hypothalamic hormones in the adenohypophysis. *Annual Review of Physiology*, **41**: 555-69.
13. Marks, N. (1978) Biotransformation and Degradation of Corticotrophins, Lipotrophins and Hypothalamic Peptides in *Frontiers in Neuroendocrinology*, Vol. 5, W. F. Ganong & L. Martini (eds), (Raven Press, New York), pp. 329-77.

14. McEwen, B. S., Krey, L. C. & Luine, V. N. (1978) Steroid hormone action in the neuroendocrine system: where is the genome involved? S. Reichlin, R. J. Baldessarini and J. B. Martin (eds), *The Hypothalamus* (Raven Press, New York), pp. 255-68.
15. Meites, J. (1980) Interactions between Hypothalamic Neurotransmitter and Pituitary Function (Symposium). Series of articles in *Federation Proceedings*, **39**: 2888-941.
16. Pearse, A. G. E. & Takor, T. (1979) Embryology of the diffuse neuroendocrine system and its relationship to the common peptides. *Federation Proceedings*, **38**: 2288-94.
17. Samuelsson, B. & Padletti, R. (eds) (1976) *Advances in Prostaglandin and Thromboxane Research*, Vol. 1, 2, pp. 1028. (Raven Press, New York).
18. Schally, A. V. (1978) Aspects of Hypothalamic Regulation of the Pituitary Gland. *Science*, **202**: 18-28.
19. Triggle, D. J. (1980) Desensitization. *Trends in Pharmacological Sciences*, 375-8.
20. Weiner, R. I. & Ganong, W. F. (1978) Role of Brain Monoamines and Histamine in Regulation of Anterior Pituitary Secretion. *Physiological Review*, **58**: 905-71.

2

NEUROENDOCRINE INVESTIGATION

2:1 Introduction

The rapid progress in neuroendocrine research over the last 25 years has resulted from the application of an increasing variety of techniques developed by anatomists, physiologists, endocrinologists, pharmacologists or biochemists. Recent approaches involve the use of several techniques in combination to tackle particularly searching problems and as such, incorporate increasingly sensitive, specific and precise analytical methods. Prior to the late 1960s, research methods in neuroendocrinology were largely the 'classical' techniques of anatomy and physiology and the results of these studies laid the foundation of neuroendocrinology as a major branch of the biological sciences. These involved mainly three areas, (a) the development of specific histological methods for studying neurosecretory hormones, (b) the perfection of surgical ablation, stimulation and lesion techniques and (c) the improvement of bioassays for hormones or neurohormones.

As the structures of the pituitary hormones, and later the hypothalamic hormones, slowly emerged as a result of the isolation and chemical characterisation studies initiated in the 1950s many bioassay methods were gradually replaced by more sensitive immunological procedures during the 1960s; particularly radioimmunoassay – the technique which revolutionised neuroendocrine research. Our present understanding of the hypothalamic control of pituitary function, and the actions of peptide or steroid hormones in general, results largely from the application of radioimmunoassay (see section 2:4(iii)) and immunohistochemical (see section 2:2(ii)) methods. Both these methods utilise the high affinity and specificity of antibodies which can be produced by appropriate immunisation of animals with synthetic hormone. During the same period of time sophisticated electrophysiological procedures were developed by the neurophysiologist to examine neural systems of neuroendocrine control mechanisms. Such procedures include antidromic stimulation techniques, microiontophoresis and more recently electrical recordings in isolated tissues (see section 2:3(ii)). Methods derived from the work of the neuropharmacologist involve the development of specific drugs which activate or block hormone receptor mechanisms and the use of radioligand binding to examine the kinetics of such mechanisms. The neurochemist's techniques have been used to examine intracellular and metabolic processes and the enzyme activities associated with hormone production, action and metabolism. All methods are constantly being refined and improved in order to answer further questions, and gain a greater understanding of the intricacies of neuroendocrine systems, and it is very likely that in the future many procedures in present use will be superseded by alternative approaches. The main methods used in neuroendocrine research will be outlined in this chapter and after each section the reader is referred to appropriate references for further details of methodology.

2:2 Histological Investigation

(i) *Early Histological Techniques*

In the first decade of this century Ramon y Cajal first described the strange-looking nerve cells whose axons projected to the posterior pituitary gland. During the 1930s histoanatomical investigations on the neurones in the hypothalamic paraventricular and supraoptic nuclei by E. & B. Scharrer and others led to the hypothesis of neurosecretion – the idea that nerve cells could have a secretory function. When the chrome alum haemotoxylin stain, first introduced by G. Gomori in 1940, was applied to the study of neurosecretory cells by Bargmann it was finally possible to demonstrate that neurones whose axons

Plate 1: Immunofluorescence Micrographs of Consecutive Mid-transverse Sections of the Rat Basal Hypothalamus (Including the Median Eminence, Arcuate Nucleus and IIIrd Ventricle – see Figure 1.2), after Incubation with Antiserum to Tyrosine Hydroxylase (TH) (A), Luteinising Hormone Releasing Hormone (B), Thyrotrophin Releasing Hormone (C), and Somatostatin (D). LHRH (B), TRH (C), and SOM (D) nerve endings are shown to be present in the median eminence in different concentrations and with different distributions. LHRH-positive nerve endings

A

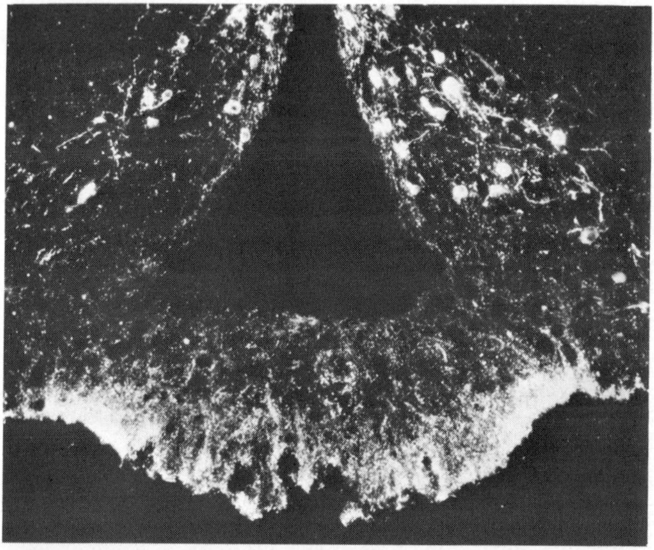

B

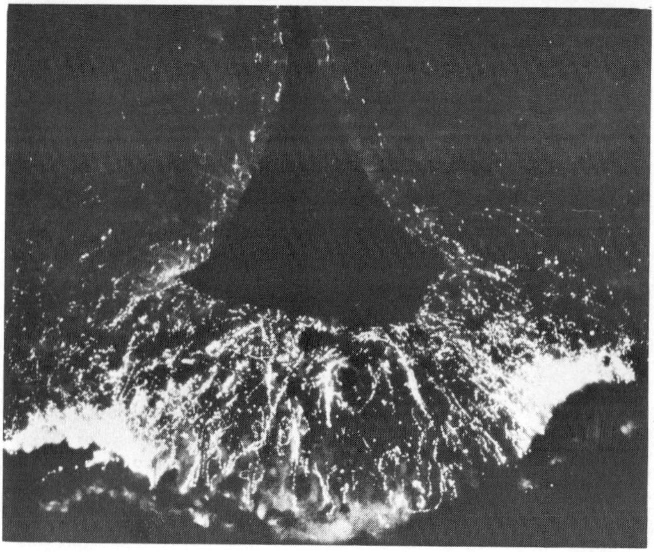

are mainly present in the lateral parts, TRH-positive nerve endings in the medial parts and SOM-positive nerve endings all over the median eminence. Note that SOM-positive nerve endings are also present in the arcuate nucleus. In this area there are also numerous TH-positive cell bodies and nerve endings mainly representing dopamine neurones (A). Magnification x75. Micrographs by kind permission of Professor T. Hökfelt, Department of Histology, Karolinska Institute, Stockholm.

C

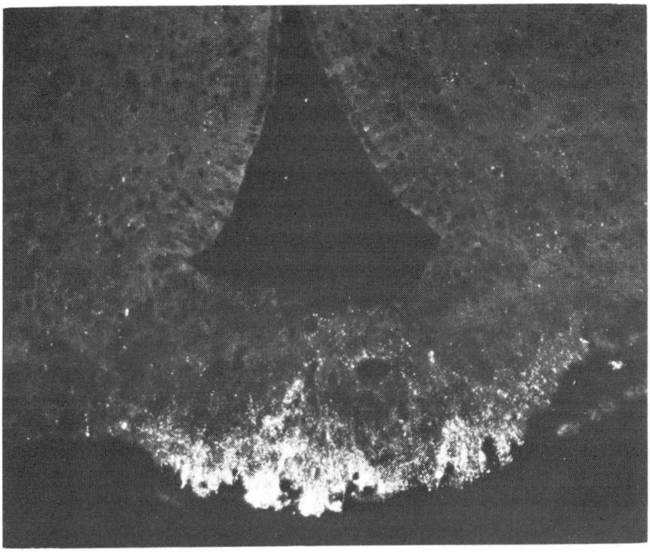

D

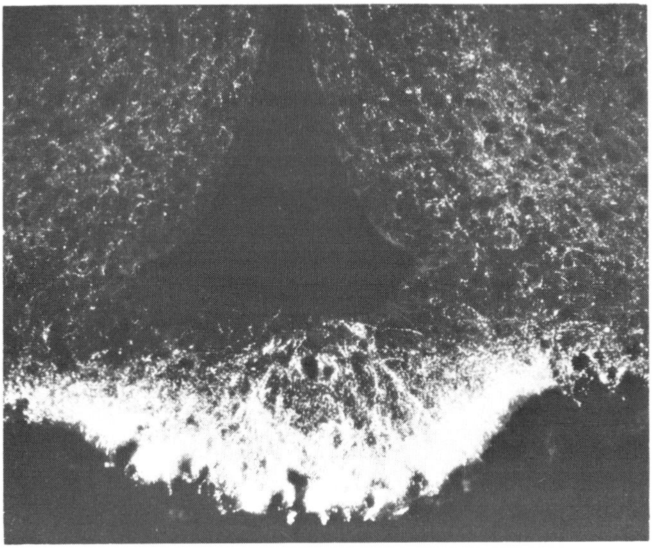

projected to the posterior pituitary synthesised and released secretory material into blood. Hence, morphological studies demonstrated the relationship between the hypothalamus and the posterior pituitary. On the other hand, elucidation of the relationship between the central nervous system and the anterior pituitary relied both on morphological studies and physiological experiments.

Early morphological investigation of the control of the anterior pituitary suggested a direct nerve supply to the gland but it soon became apparent that there was a very scant neural innervation of the anterior pituitary and that large areas were completely free from nerve fibres. In the early 1930s Popa and Fielding described the anatomy of the hypophyseal portal vessels but unfortunately ardent controversy sprang up about the direction of blood flow in them and they were disregarded for some ten years. This is surprising in view of the fact that humoral transmission of stimuli from the hypothalamus to the anterior pituitary was tentatively suggested by several workers during this period. It was not until the early 1950s that nearly twenty years of investigation culminated in the confirmation of a neuro-humoral control of the anterior pituitary gland largely as a result of the efforts of G. W. Harris and his collaborators. Most of this evidence was provided by classical physiological experiments on whole animals which encompassed such techniques as pituitary stalk section, pituitary transplantation, and experiments involving electrical stimulation or lesions of the brain. For further reading on anatomical and functional aspects of hypothalamic–pituitary systems see [2, 4, 5].

(ii) *Immunohistochemical (immunocytochemical) Methods*

In the late 1960s, immunohistochemical techniques were applied to brain and pituitary and provided a means to localise hormones, neuro-hormones or neurotransmitter systems in histological preparations (see Plate 1). The method utilises serum antibodies which, by immunisation of animals (see section 2:4(iii)), can be raised to peptide hormones, or hormone precursors or carrier proteins (e.g. neurophysin, see Chapter 3 section 4) or to neurotransmitters (e.g. 5-hydroxytryptamine) or to enzymes involved in transmitter biosynthesis (e.g. tyrosine hydroxylase for the catecholamine neurotransmitters) and other constituents. The most commonly used methods are indirect immunohistochemical techniques (see Figure 2.1). These use a second antibody, raised to the particular globulin species of the first antibody, to which a marker or label is attached. This may be fluorescent (e.g. fluorescein) so that thin slices of tissue

Figure 2.1: Indirect Immunohistochemical Techniques. (For explanation, see text.)

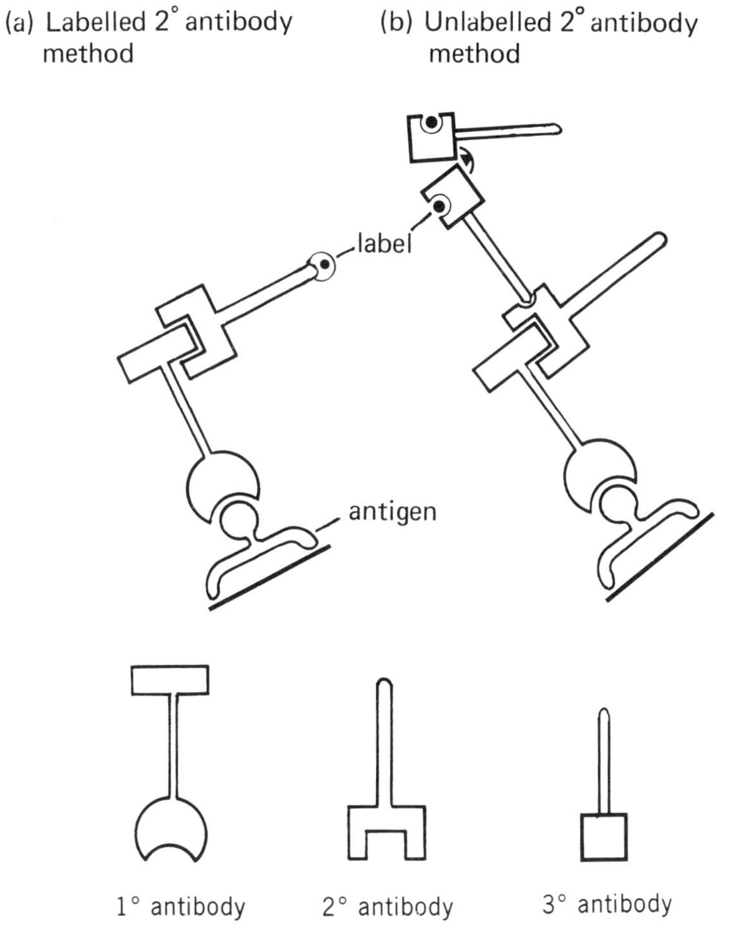

treated with both antibodies can be examined under the fluorescent microscope. Alternatively an immunoenzymatic method uses an enzyme marker (principally, horse radish peroxidase) which can be histochemically stained for visualisation under a standard phase-contrast or light microscope (see Figure 2.1(a)). The latter procedure has been recently modified to the 'unlabelled antibody' method (see Figure 2.1(b)) which uses a third antibody in the form of soluble complexes of peroxidase–antiperoxidase. This method is sufficiently precise and sophisticated to enable subcellular examination under the electron

Plate 2: Electron Micrographs of Rat Anterior Pituitary (Pars Distalis) Cells at Similar Magnification. They show the four main cell types: a somatotroph (growth-hormone secreting) magnification x6,500 (A), a corticotroph (ACTH secreting) x6,000 (B), a thyrotroph (thyroid-stimulating-hormone secreting) x6,000 (C) and a gonadotroph (luteinising-hormone/follicle-stimulating-hormone secreting) x5,200 (D). The randomly distributed cell types can be easily identified and characterised by features such as overall size and shape, size and appearance of the cell nucleus and the number of mitochondria present, but more particularly by the number, size range and distribution of the dense-core secretory granules (black). The somatotroph is a small round cell with medium-sized and large granules;

A

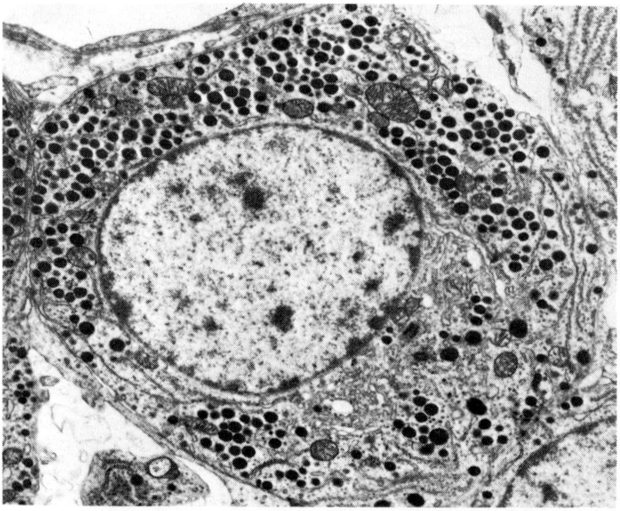

B

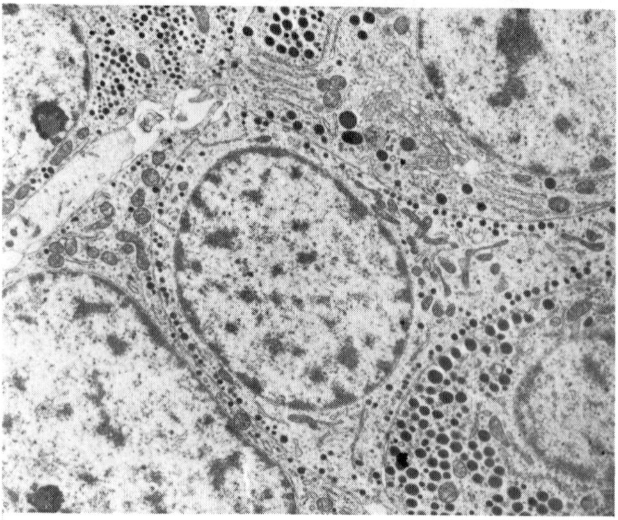

the corticotroph is an elongated, stellate cell with a single layer of small granules arranged around the periphery of the cell; the thyrotroph is similar to a corticotroph but is less elongated and has a multilayer of peripherally distributed granules; the gonadotroph is a large cell containing large numbers of small to medium-sized secretory granules. Note that a fifth type of anterior pituitary cell not shown is the lactotroph (prolactin secreting) which is generally a medium-sized cell with a few large irregularly shaped secretory granules. Micrographs kindly prepared by Mrs A. Tomlinson, Department of Human Morphology, University of Nottingham Medical School.

C

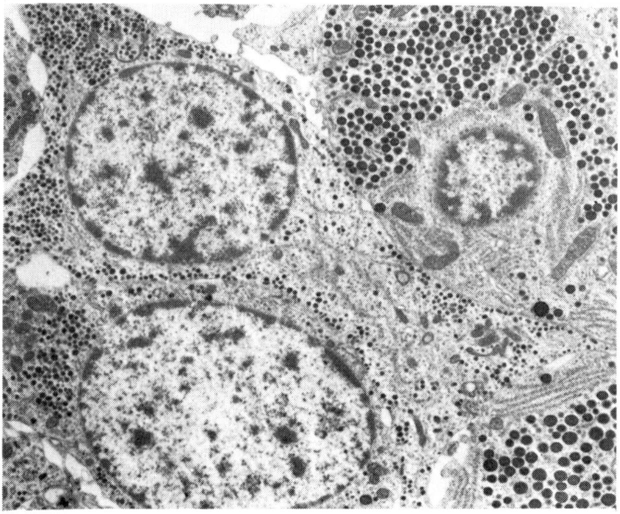

D

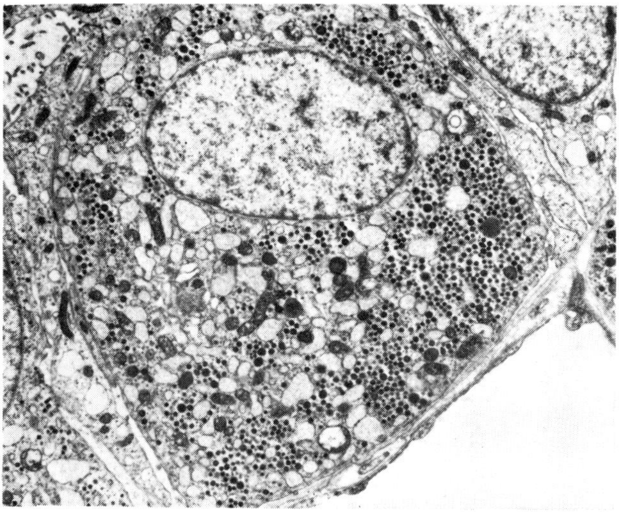

microscope. A further recent development enables the same tissue to be retreated with different antibody: in this way it has been possible to demonstrate that in particular brain regions and in certain endocrine tissues, specific cells contain both a peptide and a monoamine neurotransmitter (see Chapter 10 Section 4).

Immunohistochemical methods have provided invaluable information regarding hormone and neurotransmitter distribution (Plate 1) but there have been, and still are, many problems of specificity. Control tissue preparations are generally treated with primary antibody which has been pre-incubated with excess antigen, but often non-specific binding is common. Furthermore, the primary antibody may not be particularly well characterised as they are often in short supply, and one cannot therefore be certain of the nature of the tissue factor to which it binds; also antigen–antibody complexes may dissociate in the tissue section. A further problem with immunohistochemistry of peptidergic neurones in brain is that, although visualisation of peptide in nerve terminals and axons appears straightforward, they are not easily located in the cell bodies where they are synthesised. Reasons for this may be reduced peptide-antibody accessibility or that the peptide in the cell body is present as a precursor or prohormone which is not immunologically reactive with the antibody. One way to identify the cell bodies producing the peptide is to pre-treat the animal tissue with a drug which inhibits axonal transport (see Chapter 3 Section 3) e.g. colchicine. This sometimes causes transitory accumulation of peptide in the cell bodies to which the antibody may then bind. For further information see [1, 6].

(iii) *Electron Microscopy (EM) Studies*

This powerful tool has been widely used to examine subcellular or ultra-structures of endocrine and brain cells (see Plate 2). Its particular value has been for detailed visualisation of specifically stained or labelled material within storage granules or vesicles when used in conjugation with immunohistochemical methods or with autoradiography. For example, Gn-RH containing nerve terminals in the median eminence have been shown under the EM to make synaptic contacts with dopamine nerve-endings and this supports the evidence that dopamine regulates Gn-RH release in this region. EM studies have also enabled, for example, a detailed analysis of the incorporation and transport of ^{35}S-labelled cysteine in vasopressin–neurophysin containing granules in the axons and terminals of the neurohypophysis (see Chapter 3 Section 5). The examination of pituitary cell ultrastructure has provided information on the mechanisms of hormone synthesis, together with the identity and specific characteristics of the various cell types e.g. corticotrophs, thyrotrophs, lactotrophs, somatotrophs and gonadotrophs. Furthermore, freeze fracture EM techniques have been used to examine the synaptic membranes of neurosecretory and other neurones following secretory or release processes.

For details of practical electron microscopy see [3].

2:3 Physiological Investigation

(i) *Surgical Manipulation Techniques*

(a) *Pituitary Stalk Section.* There are many early reports on the effects of sectioning the pituitary stalk, particularly the changes which occur in the reproductive cycle of the rat. Before more appropriate assay methods were developed, ovulation or the occurrence of pseudopregnancy induced by sterile mating were frequently used as indices of anterior pituitary function. The initial results of pituitary stalk section were varied and not always easy to interpret, in that pituitary stalk section either had no effect on reproductive cycles, or resulted in prolonged disturbed cycles or led to their complete cessation. As the surgical techniques were quite difficult in the rat, these early discordant results were explained by variations in the completeness of the stalk section. Following more detailed investigations on the mammalian hypophyseal portal vessels in the late 1940s, further experiments utilising this same technique showed that rapid revascularisation (i.e. regeneration of the blood vessels) of the pituitary occurred after cutting the pituitary stalk, with consequent resumption of anterior pituitary function. However, if this regeneration was prevented by placing a plate between the cut ends, the animals remained anoestrous and their reproductive organs atrophied. Later studies demonstrated that selective transection of the portal vessels (leaving the neurohypophyseal fibres intact) impaired anterior pituitary function and collectively these experiments provided conclusive evidence for the humoral control of the adenohypophysis (see Chapter 1 Section 5).

(b) *Hypophysectomy.* Hypophysectomy, or surgical removal of the pituitary gland was successfully achieved in rats and other species as early as the 1920s using a transsphenoidal approach. Simpler techniques, which were developed later, involved removal via the external auditory canal or a parapharyngeal approach. The operation results in cessation of growth and reduction in the size and function of all endocrine glands which are under trophic control by the anterior pituitary hormones (e.g. thyroid, adrenal cortex and gonads) as well as many other metabolic disturbances. Hypophysectomy is sometimes undertaken in humans as a final resort to remove hormonal control of advanced forms of cancer (e.g. metastatic breast cancer), as adequate hormone replacement therapy with thyroid and adrenal hormones can prevent death from hypopituitarism. Hypophysectomised rats are commonly used in research to remove the source of endogenous pituitary hormones, for example for the purpose of *in vivo* bioassay methods (see Chapter 2 Section 3(iv)).

(c) *Pituitary Transplantation.* The technique of transplanting the pituitary gland to a site remote from the hypothalamus, such as under the kidney capsule, showed that the anterior pituitary secreted its hormones only if supplied by hypophyseal portal blood (see Figure 2.2). This contrasts with other endocrine glands, such as the gonads or thyroid, which can be transplanted to any part of the body and will

Figure 2.2: The Effect of Pituitary Transplantation and Retransplantation on Endocrine Function in the Rat. (Modified from the classical experiments of Nikitovitch-Winer and Everett.) The importance of the blood supply from the Median Eminence is indicated.

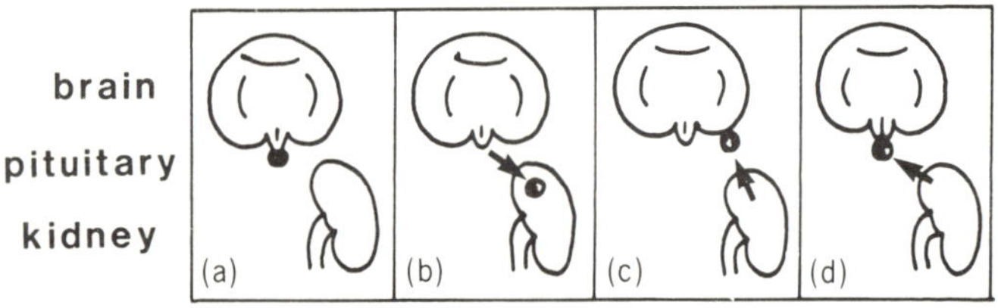

Oestrous cycles	100%	0	0	93%
Ovarian wt. (mg/100g bwt.)	27	6.1	4.3	16
Thyroid iodine uptake (24h)	7%	1.2%	.33%	3%
Adrenal wt. (mg/100g bwt.)	[illegible] 1	2.4	2.8	5.4

function normally provided they are placed where rapid systemic revascularisation can occur. Pituitary glands re-transplanted into the subarachnoid space of hypophysectomised rats (see Figure 2.2(d)) were found to maintain normal oestrous cycles whereas transplants under the temporal lobe (see Figure 2.2(c)) did not restore cyclicity even though they became vascularised by cortical and dural vessels (Figure 2.2). Similarly, thyroid and adrenal function was only restored following normal pituitary re-transplantation.

During the 1950s and early 1960s, preparations of median eminence extracts were found to contain factors which stimulated the release of corticotrophin (ACTH) and luteinising hormone and towards the end of the sixties the technique of collecting hypophyseal portal blood by cannulation of the pituitary stalk was perfected. This enabled the conclusive demonstration that only hypophyseal portal blood and not blood from the systemic circulation contained hypophysiotrophic releasing factors in concentrations that would affect anterior pituitary secretions. A number of neurohormones, neuropeptides and neurotransmitters have now been measured in hypophyseal blood (see Chapter 10 Section 5).

(ii) *Discrete Stimulation and Lesions of Brain*

(a) *Electrical Stimulation.* It has been known for some time that sensory stimuli from the environment act via a reflex pathway in the central nervous system to alter the release of anterior pituitary hormones. For example certain female animals known as reflex ovulators (e.g. rabbits and ferrets) ovulate in response to coitus, but many environmental factors influence sexual rhythms in animals and man (see Chapter 9 Section 3). As a consequence of these early ideas, Marshall and Verney, in the mid 1930s, studied the effects of applying large electric shocks through the heads of anaesthetised rabbits. They found that such diffuse stimuli evoked ovulation and induced pseudopregnancy in a large number of animals. In 1936 G. W. Harris described similar results using indirect stimulation by means of induction coils implanted in rabbits and rats. A year later he reported that localised stimulation of discrete areas of the hypothalamus also evoke ovulation. This sparked off a series of experiments in which highly localised electrical stimulation of hypothalamic and extrahypothalamic regions permitted mapping of specific brain areas influencing pituitary secretion. Stimulation of the anterior pituitary gland was shown not to evoke release of its hormones and this provided further evidence that the pathway between the hypothalamus and pituitary was not neural.

The frequency, intensity, polarity and duration of electrical stimulation and its effectiveness to stimulate hypothalamic hormone secretion has been thoroughly analysed in an attempt to understand neural activity associated with neurosecretory functions.

(b) *Stereotaxic Lesions.* At the same time that electrical stimulation techniques were being adopted by neuroendocrinologists, many workers were making converse experiments – that is placing lesions in various parts of the hypothalamus and finding diminished secretion of anterior pituitary hormones. By and large the results obtained from these two techniques, stimulation or lesions, were in general agreement. In all such studies stereotaxic techniques are employed to ensure accurate localisation of a brain region. Precise placement of stimulating electrodes or lesions is achieved in a particular animal species, such as the laboratory rat, using a stereotaxic frame and a detailed brain atlas (see Chapter 1 Section 2). The frame is designed to lock the position of the head of the anaesthetised animal so that precision probes (holding electrodes, catheters or syringe needles) may be adjusted to defined co-ordinates in each of the three dimensions using micromanipulators.

There are several methods of making localised lesions of specific neural pathways and these include:

(1) Electrolytic lesions made either with D.C. current, which will deposit ions around the site of the lesion, or with high-frequency current.

(2) Chemical lesions by which neural tissue is destroyed by local injections of toxic substances. Such substances may either be non-specific (e.g. phenol), partially specific (e.g. kainic acid which only destroys nerve cell bodies leaving adjacent axons intact), or specific (e.g. 6-hydroxydopamine which selectively destroys catecholaminergic neurones).

(3) Thermocautery or cryolesions by which tissue is destroyed by subjecting it to extreme temperatures using fine electrically heated or cooled probes.

(4) Surgical lesions in which careful microsurgical methods are used to incise specific neural pathways which impinge on the basal hypothalamus. Some of the classical surgical lesion studies were developed by a Hungarian anatomist, Halász, who used a small knife, shaped a bit like a half moon, to partially or completely deafferentate the hypothalamus. Such a deafferentated 'hypothalamic island' was found to be capable of sustaining normal pituitary function and thus these studies

led to the concept of an autonomous 'hypophysiotrophic area' in the hypothalamus. These preparations have provided much information on direct pituitary control and hormone feedback mechanisms. For further reading on the brain stimulation and lesion techniques described in this section see [7, 8, 9]

(iii) *Electrophysiological Methods*

Investigations on the electrical properties of hormone-secreting neurones are essential to the understanding of how these cells can release their hormones in response to an appropriate stimulus. Neurosecretory neurones show essentially the same properties as other neurones and the study of their electrophysiological parameters is thus an obvious complement to biochemical, morphological and immunological methods employed in current neuroendocrine research. Although electrophysiological experiments require rather elaborate electronic equipment, the recording of electrical activity within the central nervous system (CNS) is not difficult, particularly when the activity is recorded extracellularly (outside the neurone). Extracellular recordings only provide information on the frequency and pattern of firing of neurones whereas intracellular recordings will show discrete changes in resting membrane potentials and postsynaptic potentials which occur prior to the generation of action potentials. Despite the difficulty of impaling the relatively small neurones with a microelectrode, intracellular recordings have been obtained *in vivo* from neurosecretory cells in the supraoptic nucleus. More recently slices of brain tissue maintained *in vitro* have been used to investigate changes in resting potentials. Such a technique provides a potentially useful tool for investigating possible modulatory effects of hormones that may not directly affect the firing rate of neurones but may alter the resting membrane potential thus rendering a cell more or less sensitive (excitable) to synaptic inputs.

(a) *Unit Recording* (see Figure 2.3). Early electrophysiological studies relied on data obtained from macro-electrodes through which only massed potentials and multi-unit activity can be recorded. It was not until 1959 that Cross and Green presented some of the first reports on the electrical activity of single hypothalamic neurones recorded extracellularly through glass microelectrodes. Since that time electrophysiological studies have served to characterise the electrophysiology of the magnocellular neurones which terminate in the posterior pituitary (see Chapter 3 Section 8(i)) but studies on the tuberoinfundibular

Figure 2.3: Simplified Diagram Showing Electrophysiological Techniques for Antidromic Stimulation and the Ejection of Drugs on to Neurones from Multi-barrelled Microelectrodes (Microiontophoresis).

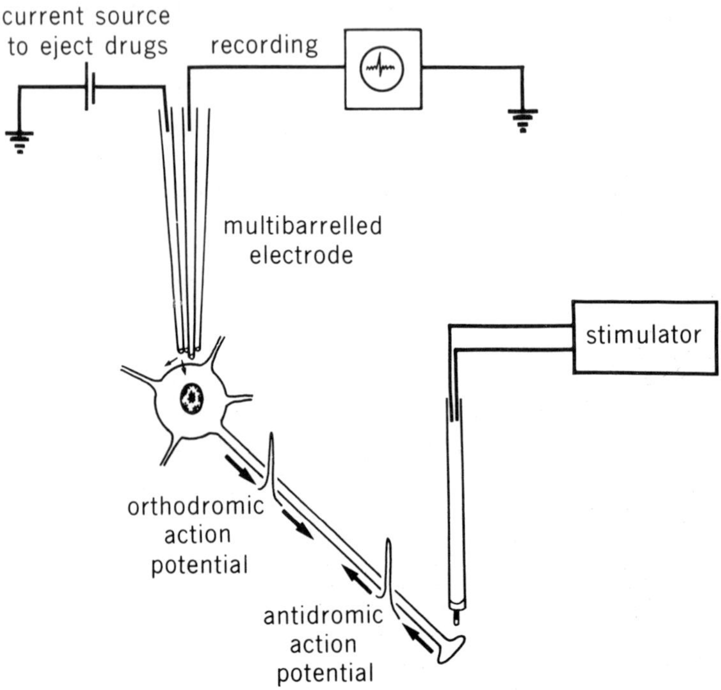

or parvicellular neurones which terminate in the median eminence and regulate anterior pituitary secretions, have not been so revealing. Initial single unit investigations were carried out on unidentified hypothalamic neurones which showed that cells in the hypothalamus changed their spontaneous firing rate in response to certain stimuli known to affect the secretion of pituitary hormones e.g. hypertonic saline, steroid hormones.

(b) *Antidromic stimulation* (see Figure 2.3). Toward the end of the 1960s and in the early 1970s the technique of antidromic activation was adopted to identify neurosecretory cells of the magnocellular system. A stimulating electrode, positioned on the pituitary stalk, is used to initiate action potentials which are conducted in a retrograde direction (toward the cell body) and can be recorded from a microelectrode positioned near the cell bodies (soma). By this means distinction can be made between cells in the hypothalamus whose axons pass through the pituitary stalk (i.e. the magnocellular neurones) and other cells which cannot be activated by antidromic stimulation and are therefore unlikely to be neurosecretory in nature.

The use of antidromic activation to identify the neuroendocrine cells of the magnocellular system has enabled characterisation of the electrical responses of these neurosecretory cells to osmotic or suckling stimuli and to establish the relationship between the patterns of unit activity and hormone secretion. The same technique has been used to identify the parvicellular neurones and antidromic invasion of action potentials can be recorded extracellularly in the hypothalamus following stimulation applied to the surface of the median eminence. For the identification of the parvicellular neurosecretory cells it is important that the stimulus is applied to the surface of the brain allowing for preferential stimulation of axons terminating on portal capillaries rather than axons coursing to the neural lobe. Unlike the antidromically identified peptidergic cells of the magnocellular system there is no electrophysiological evidence that establishes the peptidergic nature of cells which can be antidromically activated by stimulation of the median eminence. Until one can obtain electrical recordings from neurones that can be subsequently identified as neurosecretory cells by specific immunocytochemical techniques, the electrical characteristics of the tuberoinfundibular neurones cannot be firmly established.

Electrophysiological studies in the rat have shown that less than 10 per cent of recording neurones within the mediobasal hypothalamus are likely to be neurosecretory cells and have demonstrated that a substantial number of the parvicellular neurones are located in more rostral regions extending into the preoptic area. It has also been shown that some tuberoinfundibular neurones can be antidromically activated not only by stimulation of the median eminence but also by stimulation of the amygdala, thalamus, hippocampus or preoptic area. These observations suggest that axon collaterals of the neurosecretory cells also make extrahypothalamic connections with areas of the brain that provide a functional input to the hypothalamus.

The technique of antidromic stimulation has also been usefully employed to examine the possible existence of recurrent inhibition or facilitation whereby axon collaterals feedback to the parent neurone either directly by a synapse onto the neurone of origin or indirectly via local interneurones (see Figure 8.6). Recurrent inhibition results in inhibitory post-synaptic potentials in the neurosecretory cells and in extracellular recordings this would be manifested by a silent period in spontaneous activity following subthreshold stimuli for antidromic invasion. Facilitatory axon collaterals induce an excitatory post-synaptic potential and hence would increase the excitability of the neurosecretory cells. Extracellular unit studies have suggested the existence of both recurrent inhibitory and facilitatory pathways in the magnocellular and parvicellular neurones and intracellular recordings have confirmed that inhibitory post synaptic potentials occur in supra-optic neurones after pituitary stalk stimulation. The possible significance of recurrent inhibition may be as a rate limiting or feedback mechanism to control hormone secretion and may explain the phasic pattern of unit activity frequently observed in neurosecretory cells. On the other hand, facilitatory axon collaterals may be present to activate adjacent neurosecretory cells thus ensuring synchronous activation of a group of neurones and a sufficient secretion of a hormone.

(c) *Microiontophoresis.* The technique of microiontophoresis has been widely used to investigate the neuropharmacological properties of neurones. Microiontophoresis involves the use of multibarrelled glass micro-electrodes which can be filled with various concentrated drug solutions. When a current is passed through a barrel of the electrode, drug is ejected in the vicinity of a neurone from which simultaneous recordings of extracellular potentials can be obtained. In this way the response of neurones to putative neurotransmitters, hormones and other drugs may be investigated although at the present time it is difficult chemically to quantify the drugs released. The technique has been useful in establishing the sensitivity of neurosecretory cells to neurotransmitters (e.g. catecholamines, acetylcholine, 5-hydroxytryptamine) and in showing that both steroid hormones (e.g. corticosteroids, oestrogens) and peptide hormones (e.g. GnRH, TRH) can alter the rate of unit firing when applied iontophoretically. However, if a cell is shown to be responsive to a particular substance it does not necessarily imply that the cell receives synaptic inputs which release that chemical as a neurotransmitter. Thus, the physiological relevance of simply iontophoresing drugs onto neurones is open to speculation although, in general, iontophoretic studies have supported other types of pharmacological experiments investigating neurotransmitters which may be involved in controlling hormone secretion from neurosecretory cells. For further information on micro-iontophoresis techniques see [10] and for further reading on the use of electrophysiological methods in neuroendocrine research see [11, 12]

(iv) *Bioassay Techniques*

Prior to the availability of pure hormones and the introduction of radioimmunoassay techniques bioassays were widely used to estimate hormone concentrations in blood and tissues although these were frequently cumbersome and insensitive. For example the posterior pituitary hormone vasopressin was measured by its pressor activity (contraction of blood vessels) and thus relative changes in blood pressure after an intravenous infusion of vasopressin would indicate different concentrations of the hormone. Such assays, however, suffered from lack of specificity since many other hormones and factors can alter blood pressure. More specific assays were somewhat easier to develop for the anterior pituitary hormones, the gonadotrophins (LH and FSH), thyrotrophin (TSH) and corticotrophin (ACTH) because their effect on specific target glands could be determined.

Simple bioassays involved the measurement of endocrine gland weights since hyper- or hypotrophy of a gland indicated hyper- or hyposecretion of the relevant pituitary trophic hormone. Alternatively, physiological responses induced by the target glands were measured. For example, in a TSH bioassay an index of thyroid gland activity was the assessment of induced metabolic rate (see Section 6:2) by measurement of oxygen consumption. However, more precise methods were developed in which chemical techniques provided a method of measuring protein-bound iodine in the blood. Indices of cyclic ovarian function and gonadotrophin release were obtained by observing cell types in vaginal smears, and the occurrence of the preovulatory surge of LH secretion could be determined by the presence of ova in the Fallopian tubes. Ovulation is a reliable index of increased LH release and the detection of its occurrence is a simple laboratory procedure.

Reliable quantitative biological assays are now available for most known hormones, but apart from practical considerations rigorous adherence to certain fundamental criteria is necessary. It is essential to use a reliable standard preparation where inter-animal or tissue variations are a problem. As a result a number of National and International Organisations provide uniform 'international' standard preparations which are available for many hormones and should be used in order that results obtained by different laboratories may be compared. Secondly, the criteria of accuracy, precision, reproducibility, specificity and sensitivity require consideration. An accurate bioassay is one which, after sources of error have been taken into account, conforms to an accepted standard. The precision of the assay is the degree of accuracy that is feasible, or in other words, the smallest units of measurement which can be differentiated. The reproducibility of the assay assesses the extent of random errors which occur between measurements of common samples. Precision and reproducibility may be estimated by carrying out repeated determinations and then applying appropriate statistical analysis of the values. Usually 'means' and 'standard deviations' are determined. Then the term 'index of precision' may be calculated as the standard deviation divided by the slope of the log dose response and a low value, i.e. 0.2, indicates good precision. More information may be derived from the fiducial limits of error in the assay which corresponds to a given probability (often 0.95) which predicts that 95% of values fall within a particular range. Such calculations may then be used to determine whether significant differences exist between sets of values. Specificity in a bioassay refers to the determination of one physiological activity to the exclusion of others or indicates how well the assay measures what it purports to measure. Finally and most importantly, the term sensitivity defines the smallest single measurement which can be distinguished from zero with some assurance in the bioassay.

Quantitative bioassay procedures have been developed for the pituitary hormones and some of these are described in the appropriate subsequent chapters. Some bioassays are *in vivo* assays in hypophysectomised animals such as the 'tibia' test for growth hormone, but more sensitive and specific assays are usually *in vitro* assays and involve direct measurement of changes in isolated tissue preparations in response to added hormone. For example ACTH has been measured by its effect in isolated adrenal tissues, as either the reduction of ascorbic acid or the increase in corticosteroid release, measured using a spectrophotometer or a fluorimeter, or by a chemical or radioimmunoassay of steroid. Similar assays have been used for LH with isolated gonadal tissue.

With the availability of synthetic hormones, radioimmunoassay has effectively replaced many bioassays, particularly those which simply gave a qualitative index of hormone secretion without allowing any precise quantitative measurements to be made. However, because of antibodies that do not selectively detect the biologically active form of the hormone (see Chapter 2 Section 4(iii)), many bioassays are currently employed in order to confirm biological specificity. Highly sensitive cytochemical bioassays for certain pituitary hormones have been developed recently. One of the earliest of these was developed for ACTH and its sensitivity (low femtomoles – 10^{-15}M) surpasses that of available ACTH radioimmunoassays. The technique involves the measurement of redox changes in adrenal tissue slices which can be monitored by colour changes using microdensitometry. The chemically identified hypothalamic releasing hormones (TRH, Gn-RH and somatostatin) are usually measured by radioimmunoassay, but releasing factors whose structure has not been fully identified (e.g. GHRF) can be measured by their effect on pituitary hormone secretion, either *in vivo* or, more usually, by their action on pituitary tissue (glands or isolated cells – see Chapter 2 Section 3(iv)). The secretion of the pituitary hormone may be monitored either by a sensitive bioassay or radioimmunoassay.

For detailed information on specific bioassays of pituitary and hypothalamic hormones see [15] and for further reading on the biochemical investigation of the endocrine hypothalamus see [17].

(v) *Investigation of Hormone and Neurohormone Secretion*

(a) *In Vitro Techniques. In vitro* techniques are widely used to study chemical events associated with hormone secretion and to investigate mechanisms that control release of neurohormones. The maintenance of pituitary or neural tissue *in vitro* involves its immersion in a well-oxygenated iso-osmotic buffered medium. The media frequently used include Krebs bicarbonate buffer or Hanks or Lockes medium and which contain glucose as the energy substrate. Alternatively, a more specific tissue culture medium, such as Eagles, may be used which contains further appropriate nutrients. To maintain viability of the preparation an adequate supply of oxygen must be delivered and usually the medium is gassed with 95 per cent O_2 and 5 per cent CO_2. Measurement of the oxygen consumption by the tissue is indicative of intact, viable, respiring cells in the preparation and may be used to indicate suitability for secretory studies.

Two broad categories of *in vitro* systems may be defined: (1) static (fixed volume) incubations or (2) perifusion (or superfusion) systems (see Figure 2.4). With incubation methods, dissected tissue is suspended in the incubating media at a fixed concentration from which sample aliquots are withdrawn at appropriate times. Generally the effect of added test substances is compared with incubations in the absence of the substance. Released hormones can be determined in supernatant fluids following removal of the tissue by centrifugation or filtration. In this system it is important to determine optimal tissue to medium ratios. In a fixed volume of medium, hormone release is

Figure 2.4: A Typical Tissue Superfusion (Perifusion) System.

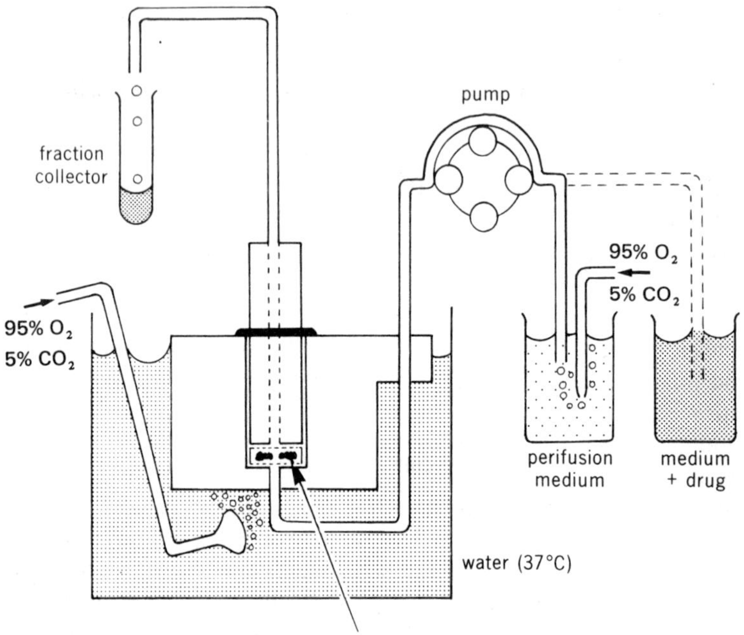

not directly proportional to the tissue concentration because of tissue metabolism of released hormone and also the effect of released hormone on the release process. In perifusion systems the tissue is constantly perifused with medium and fractions are collected from the perfusate. The latter system is often preferable for release studies in that metabolic products may be 'washed away' from the tissue and the preparation may be challenged with 'pulses' of drug solution; thus the dynamic response of the tissue can be observed using the tissue as its own control. A typical tissue superfusion system is shown in Figure 2.4.

The preparation of neural tissue for *in vitro* experiments will depend on the type of observations to be made. Intact whole or half hypothalami may be taken or the tissue may be sliced (by scalpel or slicer) or chopped (by a mechanical tissue chopper), or isolated nerve terminals (synaptosomes) may be obtained from brain homogenates subjected to sucrose density centrifugation (see Chapter 2 Section 5(i)). A method developed recently enables very accurate brain tissue dissection and this has been particularly useful for investigating regional differences of endogenous hormone levels and release within the hypothalamus. This is called the micropuncture technique in which discrete areas of tissue are 'punched' out of brain slices previously dissected from defined regions. The punches are removed using narrow bore (2-10mm diameter) needles attached to syringes which enable the punched tissue to be extruded and collected.

In vitro techniques are important for investigating the secretion of hormones from isolated pituitary glands as well as neural tissue, for example, to examine the release of hormones from the anterior pituitary gland in the absence of hypothalamic influences. Rat pituitaries can be readily dissected for appropriate incubation or tissue superfusion either by using separated neuro-intermediate lobes for *in vitro* studies on secretion of the neurohypophyseal hormones, vasopressin and oxytocin, or the intermediate lobe hormone, α-melanocyte stimulating hormone (α-MSH) and/or others, or by using separated anterior pituitary lobes. The rat anterior pituitary is flattened and readily separates into bilateral lobes. This pituitary design is very suitable for incubation studies in which one lobe can be used in the test situation leaving the other lobe as control. In this way the secretion of anterior pituitary hormones can be assessed using sensitive bioassay or radioimmunoassays and this technique has been particularly valuable in *in vitro* bioassays for the hypophysiotrophic hormones or factors e.g. CRF (see Chapter 5 Section 3(ii)). A further development of this technique which has enabled marked improvements in sensitivity has been the use of isolated pituitary cells.

Pituitary cells are isolated from anterior pituitary tissue by gentle mechanical agitation with buffered collagenase or trypsin enzyme preparations which gradually digest the connective tissue. Viable cells are dispersed in suspension after a period of time when the enzyme activity can be terminated by addition of an inhibitor. The suspensions of cells may be incubated directly or a more recent development involves the perifusion of cells on a column. The cells are carefully mixed with BioGel P2 and packed into small columns which may be perifused using a method similar to that described for undispersed tissue (see Figure 2.4). Fractions of perfusate may be assayed for hormone in the usual way. This technique has been used with cells isolated from other sources, including pars intermedia and peripheral endocrine tissues. Despite the high sensitivity of this method when used in combination with sensitive and specific assays, the isolated cell technique has some limitations for the study of receptor-mediated actions since the cell surface receptors of trypsinised cells may not respond in a normal way to hormone or neurohormone. At the present time the simplest and most direct way to investigate hypothalamic and pituitary hormone secretion is by means of *in vitro* techniques. Recently, *in vitro* brain slices have been used for multidisciplinary studies of synaptic function, in which physiological, anatomical, biochemical and pharmacological techniques are applied in concert within the same preparation. For further reading on *in vitro* release techniques see [14, 15, 17, 20].

(b) *In Vivo Techniques. In vivo* studies are those undertaken in whole animals and are obviously most important in investigating the integrated mechanisms controlling neurohormone secretion and are necessary to complement *in vitro* experiments. With appropriate sensitive bioassay or immunoassay the release of the neurohypophyseal hormones into peripheral blood may be determined directly, but assessment of the secretion of the hypophysiotrophic hormones *in vivo* is generally made indirectly, by measuring the changes in concentration of the appropriate circulating anterior pituitary hormones. Alternatively, a method has been developed for direct measurement of these hormones in hypophyseal blood (see Chapter 2 Section 3(i)). *In vivo* techniques usually require the use of general anaesthetics which creates a problem since many depress neuronal functon and can alter pituitary secretion. This may be overcome by the use of chronically implanted indwelling catheters for the use of drug injection and withdrawal of blood samples, so that the animal is allowed to recover from the effects of the anaesthetic before undertaking the study. Space does not permit description of the wide range of *in vivo* experiments which have been designed to investigate the neural control of pituitary secretions but a few of the more commonly used techniques may be listed.

The majority of experiments involve withdrawal of blood and measurement of hormone concentration either in blood serum (the cell-free component after clotting) or plasma (the cell-free component following removal of unclotted cells where heparin is used to prevent clotting). Blood can be withdrawn from indwelling catheters, usually inserted in the carotid artery, femoral artery or jugular vein, or by cardiac puncture. Trunk blood may also be obtained from animals following decapitation.

Drugs may be administered into animals either peripherally (subcutaneous –s.c., intraperitoneally – i.p. or intramuscularly – i.m. and sometimes orally) or centrally by direct micro-injection into neural tissue or into the brain ventricles (intracerebro-ventricularly –

i.c.v.). The route of administration of drugs will depend on several factors including site and duration of action, the rate of absorption into the general circulation, its half life in the blood and its rate of penetration into the brain (see Chapter 11 Section 4). When drugs are injected directly into the brain the volume of the solution is an important consideration since no more than a volume of 1μl should be injected into a rat brain in order to prevent excessive damage to neural tissue. Localised injections into discrete brain areas is a useful technique for limiting the site of action of a drug and excluding peripheral effects and this technique involves the use of a stereotaxic frame (see Chapter 2 Section 3(ii)). See [7, 9, 15] for details of specific *in vivo* methodology.

2:4 Biochemical Investigation of Hormone and Neurohormones

(i) *Chemical Extraction and Purification Methods*

Peptide or protein hormones may be isolated from tissue sources by initial extraction in a strong acid (e.g. hydrochloric acid) or acidified organic solvent (e.g. acetic acid/ethanol) using homogenisation or sonication to disrupt the tissue. Separation of the mixture is usually undertaken by various chromatographic procedures which may include gel filtration, partition or thin layer or high performance liquid chromatography (HPLC, see Section 4(iv)) and electrophoretic methods. The purification of a particular hormone depends on the availability of sensitive, specific and reliable assay methods to screen the separated fractions. Often more than one assay method is preferable in order to eliminate spurious results due to the many contaminants present. More details of the extraction methods appropriate for different hormones may be found in [15, 23].

(ii) *Chemical Identification and Characterisation*

Identification of an unknown polypeptide hormone requires a painstaking extraction and purification from large amounts of starting tissue using a reliable bioassay method. Determination of structure generally involves partial acid hydrolysis and amino acid analysis of fragments. This technique has resulted in the identification of all known pituitary hormones and certain hypothalamic hormones and neuropeptides but, owing to difficulties associated with the conditions of extraction or the bioassay, a number of others which are evident from physiological studies with partially purified hypothalamic extracts have eluded detection and identification. Identification and measurement of known hormone in blood or tissues generally involves application of radioimmunoassay or immunohistochemical methods.

(iii) *Radioimmunological Measurement of Hormones – Radioimmunoassay (RIA), radioreceptor assays*

Small molecules (including hormones) bind characteristically to large proteins often in a highly specific manner. This phenomenon is of the utmost importance in endocrine mechanisms and occurs for example, between hormones and their carrier proteins, between hormone substrates and enzymes, between hormones and receptors and between antigens and antibodies. Most hormones possess antigenic properties and have the ability to combine with antibodies (globulin proteins) which are produced by the body's immune system. This interaction was first utilised in the early 1960s to develop the technique of radioimmunoassay. Radioimmunoassay is a competitive protein-binding technique in which the hormone to be measured competes with a radioactively labelled hormone for binding sites on the protein antibody (see Figure 2.5). This technique enables the detection of small quantities of hormones in samples of extracted tissues, urine and other body fluids. It has been responsible for many major advances in endocrinology because of the high sensitivity and specificity which can be achieved. The technique has made it possible to measure basal levels of many hormones in blood and also subtle changes in their rates of secretion. Furthermore it is often more practicable than alternative bioassay methods, but does have limitations (see below). There are four basic prerequisites for setting up a radioimmunoassay:

(1) A purified preparation of hormone
(2) Radioactively labelled hormone of high specific radioactivity
(3) A specific antibody with high affinity for the hormone.
(4) A reliable technique for the separating of antibody bound and unbound hormone.

The hormone preparation to be used for radiolabelling and for unlabelled standards should be pure or in the case of large protein and glycoprotein hormones, the highest purity of extracted hormone available. This hormone should also be obtained from the same species from which the hormone samples to be measured are derived (where species differences occur). Alternatively, synthetic hormone is widely used for this purpose, but in this case the purity cannot always be relied upon since contaminating 'error peptides', resulting from the process of synthesis, may be present and these invariably possess immunoreactivity and sometimes biological activity. Labelled hormones can be

Figure 2.5: The Principles of Radioimmunoassay (RIA).

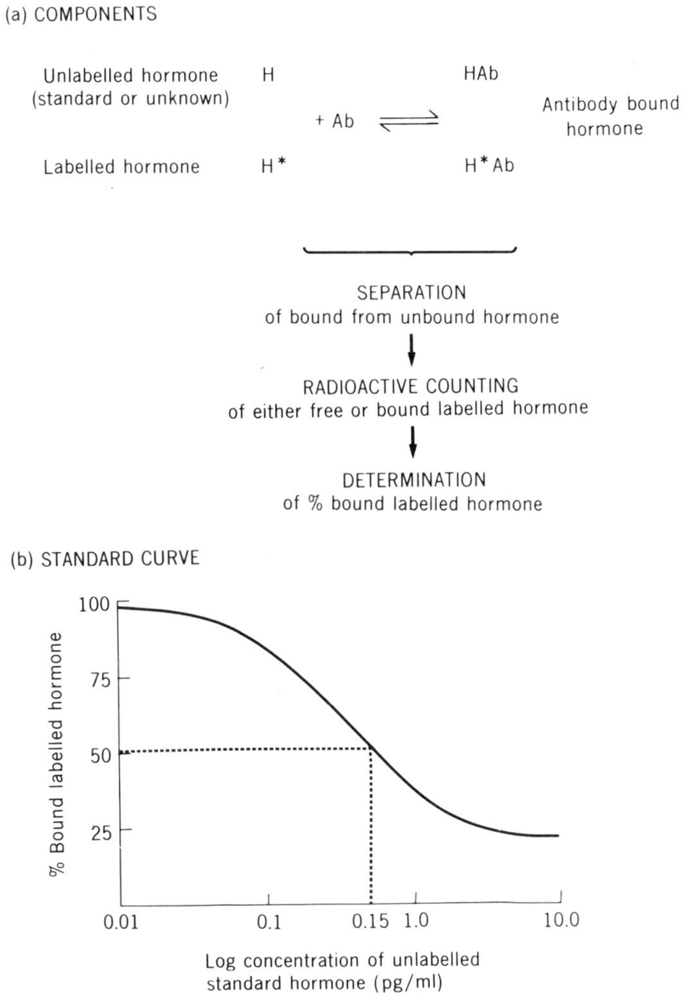

obtained from commercial suppliers or simply produced in the laboratory: such hormones are usually labelled with either radioactive iodine (^{125}I) or tritium (^{3}H) and the measurement of radioactivity will be made on either a gamma counter or a scintillation counter respectively. The labelled hormone which can be purified by gel-filtration or other form of chromatography should have a high specific radioactivity i.e. a high concentration of counts present in a low concentration of the hormone.

Antibodies to the hormone can be raised by weekly to monthly repeated injections of μg quantities of the hormone into an animal (usually rabbit or guinea pig). This induces an immune reaction and the eventual production of antibodies against the hormone. The hormones to be injected are emulsified in an oil vehicle to retard absorption into the body and then injected subcutaneously. Large polypeptide hormones such as prolactin or ACTH can be injected unchanged, whereas small molecules that are not very antigenic (e.g. steroids, TRH, Gn-RH) must be chemically conjugated to large immunogenic proteins such as albumin. Blood containing antibody is collected and stored as neat serum. The properties of a good anti-serum include (1) a high hormone binding affinity at a high dilution (1:10,000 or more), (2) an ability to displace binding of a labelled hormone with small concentrations of unlabelled hormone, and (3) a high degree of binding specificity with the particular hormone. The specificity can be checked by measuring binding or cross-reactivity with other hormones of similar structure or indeed with any other substances which are present in the samples to be assayed. In the case of peptide hormones, fragments of the molecule can be tested for cross-reactivity in order to establish the antigenic sites of the molecule which may be represented by only part of the amino acid sequence.

The general procedures for radioimmunoassays are relatively standard and involve the setting up of replicates of disposable test tubes into which the various components are accurately pipetted in small volumes. There are some variations between different assays in the buffers utilised (usually protein containing buffers at neutral pH), the conditions and duration of the antibody incubation reaction (usually at 4°C overnight) and the means of separating bound from free hormone. After incubation, when the competitive binding of labelled and unlabelled hormone with the antibody has equilibrated, it is necessary to separate bound from unbound hormone so that the percentage of labelled bound to unbound hormone may be determined (see Figure 2.5). Methods which are both reliable and easy to execute include:

(1) a second antibody, which combines with the antibody hormone

complexes forming large molecules that yield a pellet by centriguation, or
(2) a substance that will precipitate large protein antibodies but not free hormone, (e.g. alcohol, ammonium sulphate) and the resulting precipitate is separated by centrifugation, or
(3) dextran-coated charcoal, particularly suitable for small peptides, which preferentially absorbs free hormone and thus can be separated from the bound by centrifugation. The sensitivity of the assay depends on the concentration of labelled hormone used, as well as the affinity of the antibody, and this is determined by the sensitivity and efficiency of the radioactive counter. The concentration of antibody (dilution of antiserum) is selected to give 30-50 per cent binding in the absence of unlabelled hormone. A standard curve is set up by incubating various dilutions of the standard hormone preparation with the selected concentration of labelled hormone and antibody (see Figure 2.5). In the absence of any unlabelled hormone, maximum binding of the labelled hormone will occur, but as the concentration of unlabelled hormone increases, competition for the antibody displaces the labelled hormone and the amount of bound radioactivity decreases. A standard curve can then be drawn by plotting the percentage binding against the known concentration of the standards. The concentration of the hormone in the unknown samples is calculated by determining the percentage binding and then by interpolation on the standard curve (for example 50 per cent binding reads as 0.15ng/ml in Figure 2.5).

Radioimmunoassays (RIA) have been well developed for all pituitary and peripheral hormones and many have been standardised and are available in kit form, suitable for routine measurement. RIAs for all known hypothalamic neurohormones and most neuropeptides have been developed by various research laboratories. The method therefore has widespread application in both neuroendocrine research and clinical measurement and it is likely to be the assay method of choice for many hormones for a number of years to come. Nonetheless, it is a technique with a number of limitations and perhaps the most significant of these is the question of how closely the measured immunological activity relates to the biological activity of the hormones. In fact the endocrine literature abounds with examples of discrepancies between the hormone levels obtained with radioimmunological and biological assays. Antibodies are unlikely to have binding characteristics identical with those of the hormone receptor and many different antibodies with ill-defined specificity are in use. Some antibodies may preferentially cross-react with precursor or metabolised forms of the hormone to which they were raised and this may not always be rigorously tested. For this reason measurements from a radioimmunoassay are frequently checked by measurement from a bioassay so that comparisons may be made between immunological and biological activity of a hormone sample. For a more detailed analysis of the methods and problems associated with radioimmunoassay see [15, 17, 18, 24].

Clearly there are technical problems with a radioimmunoassay procedure concerning the specificity of the antibodies. In addition antibodies to a hormone which are raised in animals are not homogeneous since they are produced from multiple clones of β lymphocytes. However, a relatively new approach for raising antibodies can overcome this problem of heterogeneous antibodies. This is the technique for monoclonal antibody production by which antibodies can be obtained from a single clone of cultured cells[25]. The principle of this method involves immunising mice or rats with an antigen, using a procedure similar to that described above. A few days after the final immunisation, the animal's spleen contains many lymphocytes which will be making antibodies to the appropriate antigen. Spleen cells are then mixed with selected stock myeloma cells in the presence of polyethylene glycol. This results in fusion of cells and the production of hybrid myeloma/spleen cells. The cells are then cultured in a specific medium known as 'HAT medium', which enables the hybrid cells to survive but causes the unhybridised myeloma cells to die because they are a mutant strain lacking an important enzyme utilised in the salvage pathway of nuclei acid metabolism. Unfused spleen cells die naturally in culture. In this way only the hybridised cells are selected, some of which (typically 10%) will be producing antibodies to the antigen. The medium is then tested for its ability to bind a labelled antigen (say hormone) indicating the presence of antibody secreting hybrid cells. The cells are then cloned. This is done either by diluting the cells to near 1 per culture and growing them to a high concentration, or by plating them out on soft agarose until individual cells have multiplied sufficiently so that a colony can be picked up with a micropipette and grown in further medium. Thus a cell line, developed from a single hybrid molecule, can be established which will produce homogeneous antibodies to the antigen. Some advantages of monoclonal antibodies over conventional antisera is that their binding properties are identical and thus have exceptional specificity, they are insignificantly contaminated with other antibodies, and they can readily be labelled simply by growing the cells in a medium containing radioactively labelled amino acids. Perhaps one of the most important advantages of monoclonal antibodies is that they can be reproduced indefinitely and thus radioimmunoassays and related techniques could be standardised on an international basis. This specialised technique is now gaining wider application for the production of antibodies to hormones and may eventually supersede the conventional methods generally employed by endocrinologists.

An alternative competitive protein binding assay procedure for the measurement of hormones, which also involves the use of radiolabelled compounds, is the radioreceptor assay. The principles of this assay are similar to those of radioimmunoassay described above, but instead of measuring hormone binding to an antibody, hormone binding to a receptor preparation is assessed. Tissue (or blood), which contains high concentrations of the appropriate receptor, is homogenised and purified so that a stable preparation with a high density of receptors is obtained. This preparation is then used to measure the receptor binding of a labelled compound in the presence of unlabelled hormone (a standard or sample). Similar to the radioimmunoassay technique a standard curve is constructed and unknown samples are read off the curve. These assays have, to date, been of limited application because of the difficulties associated with the preparation and long-term stability of receptor proteins although they have been particularly important for determining plasma drug concentrations.

Most steroid and the thyroid hormones are normally transferred in blood bound to specific plasma globulins – e.g. transcortin (for cortisol), sex hormone binding globulin, and thyroxine binding globulin – and such proteins have been used in competitive protein binding assays with radio-isotopically labelled hormone. However, most assays, including radioimmunoassays, for these hormones measure total hormone activity (i.e. bound and unbound) although more specialised methods are available for the measurement of free thyroid hormone for example. If the binding globulin concentration in the blood sample is altered then measurement of total hormone activity is misleading since it is only unbound hormone which is physiologically active and this active fraction may remain within the normal range when binding globulin concentration is high. These factors are clearly of importance in clinical measurements; for example women on the contraceptive pill may have raised thyroxine binding globulin and thus high total thyroxine (T_4) due to the elevated oestrogens, but their free T_4 may be unchanged.

Figure 2.6: High-Performance Liquid Chromatography (HPLC) — Separation and Analysis of Hormones, Neuropeptides and Neurotransmitters.

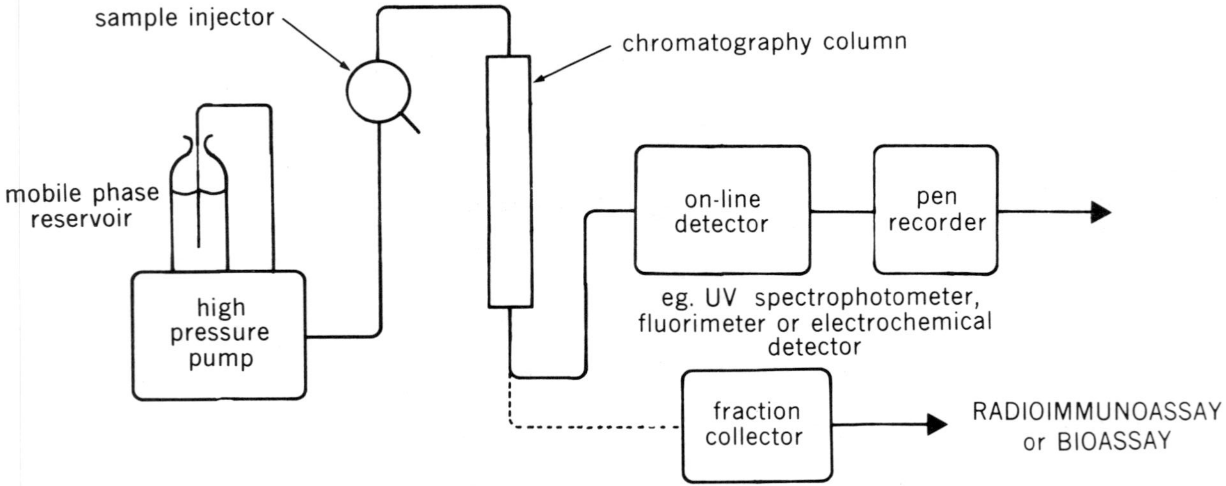

(iv) *High-Performance Liquid Chromatography (HPLC)*

With respect to the measurement of peptide hormones and neuropeptides, a particular problem alluded to in the previous section is that of distinguishing between the various forms of closely related peptide (i.e. prohormone precursors, intact hormone or metabolised peptide fragments) which may be present in blood and tissue samples (see Chapter 11 Section 4). The obvious approach is to separate out the components prior to assay. This is not a particularly straightforward procedure for small samples but the recently developed technique of HPLC is now enjoying widespread application. The principle is based on either the hydrophobic (non-polar) or electrical charge (polar) differences between similar peptides and the procedure is to separate the components by injecting the sample rapidly through a specialised chromatography column by means of a high-pressure pump (see Figure 2.6). The most common type of column is called reverse-phase and consists of a non-polar treated silica material. The sample is washed through with a mixed buffer–organic solvent mobile phase so that the peptide components are separated according to their hydrophobic properties. Alternative techniques use ion-exchange columns which separate the peptide components according to their charge properties. The separated components may be collected in a fraction collector for measurement by radioimmunoassay or other method. Alternatively various forms of on-line detector may be used at the column effluent (see Figure 2.6) such as a U.V. spectrophotometer, a fluorimeter or an electrochemical detector. Further details of the application of this rapidly advancing technique to peptide hormones and neuropeptides are described in papers listed in [22].

(v) *Isotopic Tracer Methods and Autoradiography*

Radioactive isotopes are not only used in radioimmunological assay procedures, but they have been indispensible in isotopic tracer methods. The technique consists of incorporating the isotope into living tissue usually as part of a precursor or enzyme substrate or hormone pool and taking the tissue at various times after introducing the isotope. This makes it possible to follow the fate of that substance *in vivo* under conditions which cause minimum disturbance to the normal activity of the tissue. The choice of isotope depends on the chemical nature of the substance under study and also the isotope's rate of decomposition, usually expressed as half-life (time required for the isotope to halve its activity). Radioactive isotopes commonly used include carbon (^{14}C), tritium (^{3}H), sulphur (^{35}S), iodine (^{125}I or ^{131}I) whose half lives range from 5000 years to 8 days. Isotopic tracer methods have been critical in the determination of metabolic pathways in cells in general but also have particular endocrine applications. For example, the determination of hormone receptor sites by binding studies with labelled hormone. This method has provided the basis for our knowledge of specific hormone receptor tissues particularly when used in conjunction with the technique of autoradiography as follows. After injection or treatment with an isotope, histological sections of the tissue are left in contact with a photographic plate or film for an appropriate time to develop an image due to the isotope present. The sections can be subsequently stained and thus cellular localisation of the bound labelled hormone can be visualised. Labelled substances can also be separated on paper or thin-layer chromatography and be similarly located by autoradiography. These techniques have been particularly valuable in determining the location of target cells, including brain areas, which bind steroid hormones (e.g. oestrogen and corticosteroids, see Chapter 1 Section 11). Another neuroendocrine area which has made use of isotopic tracer methods is the study of the synthesis, transport and storage of neurohormones in hypothalamic neurosecretory neurones. For example the technique has provided detailed information on the processing of the neurohypophyseal hormones vasopressin, oxytocin and their respective neurophysin carrier proteins by tracing the transportation of incorporated $[^{35}S]$ cysteine and other isotopes (see Chapter 3 Section 3). For further practical details see [3B].

(vi) *Nucleotide Tracer Techniques*

Techniques developed by molecular biologists have been used recently to study the localisation and synthesis of peptide hormones (see [19]). This involves making specific complementary DNAs (cDNAs) to probe for messenger RNAs (mRNAs) for the peptide hormones. If cells are making the hormones the cells will of course contain these mRNAs and thus the cDNAs will bind with them. When the cells synthesise a peptide which is not identical to but resembles the true hormone, the cDNAs will also bind with the mRNAs but not so tightly. It is then possible to extract the bound labelled cDNAs and mRNAs and to decode the mRNAs. This decoding will thus determine precisely what peptides are being synthesised in the cell, and distinction can then be made between cells that synthesise a known peptide hormone and those which synthesise structurally related molecules. Immunological techniques may not show such differences due to cross-reactivity with the antibody. These cDNA probes may also be used to determine the sequences of hormone precursors (pro-hormones) and to elucidate

See Hardy + Kronenberg in "Recent Advances in Endocr + Metab." 2 (1982)

which specific hormones are processed from the common precursors in different cells. The application of molecular biology techniques to study the central nervous system and brain function is developing rapidly and within the next few years it is likely that our understanding of such specific protein synthesising mechanisms in brain will be improved considerably.

2:5 Neurochemical and Pharmacological Investigation

A number of neurochemical and pharmacological procedures are used to study the neurohormones and neurotransmitters which regulate pituitary function. Unfortunately in the present book, space does not allow detailed descriptions of these procedures and we shall simply outline some of them. Sources of further reference to these methods for the interested reader may be found at the end of this section.

(i) *Studies with Isolated Nerve-ending Preparations (Synaptosomes) and Synaptic Vesicles*

Neurochemical studies of the mammalian hypothalamus and other brain regions require isolated preparations because of the morphological and functional complexity of these neural tissues. It is difficult to isolate undamaged neurones free from surrounding glial and other cells, but it is possible to isolate a functionally important region of the neurone – the pre-synaptic bulb or nerve-ending – using controlled tissue homogenisation and gradient ultracentrifugation techniques to purify the subcellular elements (see [14, 21] for details). The nerve endings (synaptosomes) are formed during homogenisation by rapid resealing of their membranes where they are sheared away from the axons, and survive intact as miniature enucleated cellular particles containing mitochondria, synaptic vesicles and cytosol. Synaptosomes are metabolically and electrically active preparations and have been widely used to study mechanisms of neurotransmitter and neurohormone release and interaction. The release of TRH, Gn-RH, somatostatin, CRF and other peptides from synaptosomes isolated from specific brain regions, particularly the hypothalamic median eminence, can be studied and a release mechanism dependent on the presence of free calcium ions has been demonstrated. A similar release of vasopressin from neurohypophyseal synaptosomes has also been studied.

High concentrations of various neurotransmitter amines are present in the perivascular zone of the hypothalamic median eminence where peptidergic nerve terminals are close to capillaries of the hypophyseal portal vessels (see Chapter 1 Section 4). There is evidence that neurotransmitter released in the median eminence may modulate the secretion of the neuropeptide hormones which regulate anterior pituitary function (see Figure 1.8). Morphological evidence for axo-axonic connections between dopaminergic and peptidergic terminals in this region has been described but studies with synaptosomes isolated from the hypothalamus of freshly killed animals have provided the most convincing demonstration for functional relationships between neurotransmitters and neuropeptide hormones in the median eminence. For example, histamine stimulates the release of TRH but not other peptides and dopamine stimulates the release of TRH and LHRH although its effect on somatostatin release is inhibitory (see [14] for further reading).

(ii) *Studies with Receptor Preparations, Ligand, Agonists and Antagonists*

Much of this book is concerned with the interaction of peptide hormones, neurohormones, neurotransmitters and drugs with specific cells to induce particular biological effects and as described previously in this chapter, there are various methods of studying these effects using whole animals or isolated tissue preparations. It is well substantiated that such chemical messengers or drugs bind as ligands to a receptor substance which usually resides in the target cell membrane (see Chapter 1 Section 10). In order to examine the precise mechanisms of ligand–receptor interactions, a method is required which circumvents the effects of other processes in tissues, such as transport, distribution and metabolism. This is possible with ligand binding in cell-free 'receptor' preparations, and the method has become feasible with the development of ligands (hormone transmitter or drug) with both a high specificity and affinity for the receptors. These ligands may be agonists (which activate or stimulate the receptor and effector mechanisms) or more usually, antagonists (which block these mechanisms but still bind to the receptor with high affinity).

Receptor preparations are partially purified subcellular fractions isolated by ultracentrifugation. Various concentrations of radioligand are added to aliquots of the preparation and the bound ligand is separated from the unbound ligand by centrifugation or filtration. However, a proportion of the binding is not specific for the receptor since the ligand will bind to other components in the preparation (non-specific binding). In order to determine the magnitude of non-specific binding an unlabelled compound, which is also very specific for the receptors, is added to the receptor preparation at a high concentration which will prevent the high-affinity radioligand binding to the receptor but not the low-affinity non-specific binding. By subtracting the amount of ligand bound in the presence of this compound from the total binding the specific receptor binding can be determined. Using different concentrations of the radioligand the maximum binding (i.e. receptor density) and receptor affinity can be calculated. In this way quantitative measurements of receptor characteristics can be made and changes of receptor numbers and/or affinity which occur in different physiological conditions can be studied.

The interaction and relationship between the receptors and their primary effectors (very often adenylate cyclase in endocrine cell membranes) can also be investigated using similar preparations, as can the factors which regulate receptor synthesis, internalisation and degradation (see Chapter 1 Section 10). Techniques for further purification and isolation of receptors has been undertaken in a number of cases using biochemical procedures similar to those used in the isolation and characterisation of hormones and neurohormones (see Chapter 2 Section 4(i)). For details of methodology of receptor techniques see [9, 20].

(iii) *Studies with Enzyme Preparations and Biotransformation Processes*

Enzymes of particular interest to the neuroendocrinologist are proteolytic enzymes (peptidases) which cleave peptide bonds. These enzymes firstly process active peptide hormones from precursor forms and secondly metabolise these hormones into inactive (or further active) fragments (Chapter 3 Section 3). Degradative peptidase enzymes in particular are widely distributed in tissues and may play an important role in controlling hormone action. These peptidases not only inactivate hormone in blood or remove hormone in the liver and kidneys, but also act at the sites of hormone production to regulate hormone release and receptor interaction processes. The availability of radioactively labelled synthetic peptides and the development of sensitive and specific radioimmunoassays have enabled the study of these biotransformation processes for the hypothalamic regulatory hormones and other neuropeptides using dispersed (e.g. homogenised) preparations and subcellular fractions isolated from the hypothalamus, other brain regions and pituitary. These techniques may involve quantifying enzymic degradation by monitoring the disappearance of added hormone or, better still, the separated metabolites or fragments produced are measured. The latter procedure may employ high resolution HPLC methods. For further reading see [3, 16] and [13] in Chapter 1.

The study of hormone biosynthesis and prohormone processing involves alternative approaches and these are discussed for the neurohypophyseal hormones in the next chapter and for the corticotrophin family of peptides (see Chapter 5 Section 2). More recent developments for the study of the biosynthesis of the hypothalamic regulatory hormones include rigorous procedures of biochemical isolation and purification of precursor forms coupled with HPLC and radioimmunoassay with various highly characterised antibodies, or alternatively the use of methods of molecular biology to isolate and study the appropriate nucleotides (DNA or RNA) for the transcription and translation of these precursor proteins. Furthermore, the biosynthesis of neuropeptide hormones frequently involves post-translational enzyme changes to introduce blocked *N*- and *C*-terminal amino acid residues (e.g. *N*-pyro-Glu or *C*-terminal amides in TRH and Gn-RH; or *N*-acetyl in α-MSH; see Chapter 4 Section 2(i)) and these require further specialised approaches. For further reading on methods used to study hormone and neuropeptide biosynthesis see [8, 12] in Chapter 3, [5] in Chapter 5, and [11] in Chapter 8.

In addition to the indirect studies of enzyme antivity, techniques are also used to isolate and purify the enzymes concerned (see [13]). The methods for isolation of active enzymes naturally differ from those used to isolate the peptide hormones where acid-based extraction is undertaken to prevent or destroy enzymic activity. Usually the tissue, which may be a specific subcellular component, is gently dispersed in isotonic aqueous solutions. In order to release enzymes which may be 'locked' in membranes or subcellular particles, agents such as non-ionic detergents (e.g. Triton *X*-100) or lipid solvents may be used. Often enzymes are initially extracted from neural tissues by the use of cold acetone which is then dried to a powder in the absence of water. Further purification may then occur after resuspension in appropriate solvent or buffer using techniques of electrophoresis and affinity chromatography. The purpose of purifying the enzymes is to examine their properties and kinetic activity with substrate(s) in cell-free systems and this can provide valuable information, since enzyme abnormalities may be responsible for dysfunction or disease. Kinetic studies of neural enzymes is a specialised technique and generalised accounts for the interested reader may be obtained in most biochemistry text books or more specific reviews listed below. For further information on the methods of study of brain enzymes see [21].

References

(i) *Anatomy/Morphology*

1. Elde, R., Coulter, H. D. & Seybold, V. (1980) Immunohistochemical Techniques for the Analysis of Peptidergic Neurones, in J. L. Barker & T. G. Smith (eds), *The Role of Peptides in Neuronal Function* (Marcel Dekker Inc., New York), pp. 1-20.
2. Everett, J. W. (1978) The Mammalian Hypothalamo-Hypophyseal System, in S. L. Jeffcoate & J. S. M. Hutchinson (eds), *The Endocrine Hypothalamus* (Academic Press, New York), pp. 1-33.
3. Glauert, A. M. (ed.) (1974) A. *Practical Methods in Electron Microscopy* (1977). B. *Autoradiography and Immunocytochemistry*. Vol. 1-3. North Holland Publ.
4. Heller, H. & Lederis, K. (eds) (1971) *Subcellular Organisation and Function in Endocrine Tissues*. Cambridge University Press.
5. Joseph, S. A. & Knigge, K. M. (1978) The Endocrine Hypothalamus: Recent Anatomical Studies, in S. Reichlin, R. J. Baldessarini & J. B. Martin (eds), *The Hypothalamus* (Raven Press, New York), pp. 15-67.
6. Larsson, L.-I. (1981) Immunocytochemistry of Secretory Peptides: Immunochemical dissections of Peptide Diversity. *Journal of Histochemistry and Cytochemistry*, **29**: 1032-42.

(ii) *Physiology/Pharmacology*

7. Bradley, P. B. (ed.) (1975) *Methods in Brain Research*. John Wiley & Sons.
8. Cooper, J. R., Bloom, F. E. & Roth, R. H. (1978) *The Biochemical Basis of Neuropharmacology*. Oxford University Press, New York.
9. Iverson, S., Iverson, L. L. & Synder, S. M. (1982) *Handbook of Psychopharmacology*.
10. Kozhechkin, S. N. (1980) Microiotophoretic Techniques in Pharmacology. *Trends in Pharmacological Sciences*, **1**: 185-7.
11. Lincoln, D. W. (1978) Investigation of Hypothalamic Function: Anatomical and physiological Studies, in S. L. Jeffcoate & J. S. M. Hutchinson (eds), *The Endocrine Hypothalamus* (Academic Press, New York), pp. 35-74.
12. Renaud, L. P. (1978) Neurophysiological Organization of the Endocrine Hypothalamus, in S. Reichlin, R. J. Baldessarini & J. B. Martin (eds), *The Hypothalamus* (Raven Press, New York), pp. 269-300.

(iii) *Biochemistry/Neurochemistry*

13. Bauer, K., Knisatschek, H., Horsthemke, B. & Sievers, J. (1981) Proteolytic Enzymes and Neuropeptide Metabolism, in R. P. Millar (ed.) *Neuropeptides: Biochemical and Physiological Studies* (Churchill Livingstone, Edinburgh), pp. 161-70.
14. Bennett, G. W. (1981) Release of neuropeptide hormones from synaptosomes: response to depolarization and neurotransmitters, in R. P. Millar (ed.) *Neuropeptides: Biochemical and Physiological Studies* (Churchill Livingstone, Edinburgh), pp. 61-77.
15. Berson, S. A. & Yalow, R. S. (eds) (1973) *Methods in Investigative and Diagnostic Endocrinology* (North Holland Publ.).
16. Griffiths, E. C. & Kelly, J. A. (1979) Mechanisms of Inactivation of Hypothalamic Regulatory Peptides. *Molecular and Cellular Endocrinology*, **14**: 3-17.
17. Jeffcoate, S. L. (1978) The Biochemical Investigation of the Endocrine

Hypothalamus, in S. L. Jeffcoate & J. S. M. Hutchinson (eds) *The Endocrine Hypothalamus* (Academic Press, New York), pp. 107-48.

18. Kirkham, K. E. & Hunter, W. M. (eds) (1971) *Radioimmunoassay Methods* (Churchill Livingstone, Edinburgh).
19. Kolata, G. (1982) Molecular Biology of Brain Hormones. *Science*, **215**: 1223-4.
20. Lynch, G. & Schubert, P. (1980) The Use of *in vitro* Brain Slices for Multidisciplinary Studies of Synaptic Function. *Annual Review of Neuroscience*, **3**: 1-22.
21. McIlwain, H. (ed.) (1975) *Practical Neurochemistry* (Churchill Livingstone, Edinburgh).
22. Molnar, I. & Horvath, C. (1977) Nice, E. C., Capp, M. & O'Hare, M. J. (1979) Desiderio, D. M., Stein, J. L., Cunningham, M. D. & Sabbatini, J. Z. (1980) McDermott, J. R., Smith, A. I., Biggins, J. A., Al-Noaemi, M. C. & Edwardson, J. A. (1981) *All* in *Journal of Chromotography*, **142**: 623-40, **185**: 413-27, **193**: 369-77, **222**: 371-9.
23. Sandow, J. & Konig, W. (1978) Chemistry of the Hypothalamic Hormones, in S. L. Jeffcoate & J. S. M. Hutchinson (eds) *The Endocrine Hypothalamus* (Academic Press, New York), pp. 150-211.
24. Yalow, R. S. (1980) Radioimmunoassay: a Major Advance but not without Problems. *Trends in Pharmacological Sciences*, **1**: 266-8.
25. Yelton, D. E. & Scharff, M. D. (1981) Monoclonal Antibodies: a Powerful New Tool in Biology and Medicine. *Annual Review of Biochemistry*, **50**: 657-80.

3

CONTROL OF THE POSTERIOR PITUITARY (NEUROHYPOPHYSIS)

3:1 Hormones of the Neurohypophysis **3:2** The Hypothalamic Neurohypophyseal System **3:3** Biosynthesis and Axonal Transport of Neurohypophyseal Hormones **3:4** Neurophysins (i) Hormone carrier proteins (ii) Neurophysin localisation (iii) Biochemical studies **3:5** Storage of the Neurohypophyseal Hormones **3:6** Neurosecretion of Neurohypophyseal Hormones (i) Stimulus for secretion (ii) Mechanism of hormone discharge (exocytosis) (iii) Role of calcium ions in neurosecretion **3:7** Physiological Control of Neurohypophyseal Hormone Secretion (i) Oxytocin secretion (ii) Vasopressin secretion **3:8** Electrophysiological Studies of the Magnocellular Neurosecretory Cells (i) Magnocellular neurones of the neurohypophyseal system (ii) Oxytocinergic neurones (iii) Vasopressin neurones **3:9** Neurotransmitters and Neuropeptides (i) Neurotransmitter regulation of the magnocellular neurones (ii) Neurotransmitters and neuropeptides in the Posterior Pituitary **3:10** Actions of the Neurohypophyseal Hormones (i) Oxytocin and vasopressin (ii) Roles of secreted neurophysins

The structures of the neurohypophyseal hormones and their site and mode of production and secretion was outlined in Chapter 1 (Section 6). Here we shall consider in more detail the control of posterior pituitary hormone secretion and the experimental evidence which has formed the basis for our understanding of this system.

3:1 Hormones of the Neurohypophysis

The physiological importance of the neurohypophysis has been known since the beginning of the century. By the 1930s the existence of four separate biological activities had been shown, i.e. pressor (increase in blood pressure), antidiuretic (reduction in urine excretion), oxytocic (contraction of uterine muscle) and galactogogic (contraction of smooth muscle in the mammary gland causing milk ejection). At the same time, the concept of neurosecretion of protein-like material from neurones was developing, largely due to the histological studies of Ernst Sharrer working with fish and other lower vertebrates. It was following the introduction of the chrome-haematoxylin (Gomori) staining

Figure 3.1: Structure of Mammalian Posterior Pituitary Hormones.

	S———————————————S (disulfide bridge, 1–6)									
COMMON STRUCTURE –	CYS	TYR –	✳ –	GLU –	ASN –	CYS –	PRO –	✳ –	GLY –	NH_2
(ONLY AMINO ACIDS IN POSITIONS 3 AND 8 DIFFER)	1	2	3	4	5	6	7	8	9	
ARGININE VASOPRESSIN (AVP, ADH)			PHE					ARG		
LYSINE VASOPRESSIN (LVP) – PIG FAMILY			PHE					LYS		
OXYTOCIN			ILE					LEU		
ARGININE VASOTOCIN (AVT) PRESENT IN FOETAL MAMMAL?			ILE					ARG		

technique in the 1940s that Bargman and colleagues established in the early 1950s that neurosecretory cells in specific regions of the mammalian hypothalamus (the supraoptic and paraventricular nuclei) synthesise a 'neuro-secretory substance' which is transported down the neural axons to the posterior lobe of the pituitary where it is stored and eventually released.

The description of a hypothalamic–neurohypophyseal secretory system was rapidly followed by biochemical characterisation of the hormone involved. After some debate as to the number and molecular size of the neurosecretory hormones, Du Vigneaud and his colleagues determined the structures of oxytocin and vasopressin (Figure 3.1) as early as 1953 using techniques of enzyme degradation and chromatographic analysis. The same techniques had been used successfully by Sanger to determine the structure of the polypeptide hormone, insulin. The use of acetic acid to extract these hormones from the neurohypophysis was particularly pertinent for the isolation of low-molecular-weight peptide hormones and their separation from the larger carrier proteins, the neurophysins, with which they were later shown to be associated. It rapidly emerged that oxytocin stimulated milk ejection and uterine contractions and vasopressin was antidiuretic as well as possessing pressor activity. Thus the presence of two hormones was able to account for the four activities shown in neurohypophyseal extracts.

3:2 The Hypothalamic Neurohypophyseal System

The mammalian hypothalamic neurohypophyseal system (also termed magnocellular system) consists of large, multipolar cells in the paired supraoptic (SON) and paraventricular nuclei (PVN) and the internuclear cells which form a discontinuous bridge between the PVN and SON (see Figure 1.2). The majority of nerve fibres originating from the PVN arch ventrally toward the SON where they join the outflow of the SON. Together the axons run medially in the floor of the hypothalamus, (the supraopticohypophyseal tract) to terminate on neurohypophyseal capillaries in the pituitary stalk and in the posterior lobe or neurohypophysis. Neurones from these nuclei also project to the zona externa of the median eminence where the nerve endings terminate in close proximity to the hypophyseal portal capillaries. A third projection of this system is to the cerebrospinal fluid (CSF) of the third ventricle. These latter two projections were demonstrated histologically in the 1950s by Bargman & Scharrer but were largely ignored until the system was re-examined with more sensitive immunohistochemical techniques using neurophysin as the tissue marker. Quantitatively the major site of release of oxytocin and vasopressin is the neurohypophysis although the two minor pathways may also subserve important physiological functions.

Very recently, a further specific projection from the PVN to the lower brain stem and spinal cord has been identified by the immunocytochemical method (see Chapter 2 Section 2(ii)). This projection is chiefly oxytocinergic, but somatostatin and dopamine containing neurones have also been implicated. Identification of this projection has created discussion concerning the PVN as an important centre for integration of autonomic and neuroendocrine responses. Immunoreactive vasopressin and oxytocin and their neurophysins are found in both the SON and PVN, although oxytocin is largely associated with the PVN and vasopressin with the SON. However, within each nucleus regional differences in the prevalence of the two hormones have been found in several species. In the mouse and rat, vasopressin and its specific neurophysin is also found in the suprachiasmatic nuclei (see Chapter 9 Section 2) but in the human brain no clear nuclei can be recognised in this area. For further reading on the anatomical aspects of the magnocellular neurosecretory system see [3].

3:3 Biosynthesis and Axonal Transport of Neurohypophyseal Hormones

Research studies in the 1950s showed that the 'neurosecretory material' (identified in secretory granules by histological methods) and the 'hormone activity' (measured by bioassay) in the neurohypophyseal system of rats and other mammalian species was always synthesised and transported together, following an appropriate stimulus. Experiments also demonstrated that when the axons of the neurohypophyseal cells were interrupted or lesioned, neurosecretory material and hormone activity accumulated on the cell body side of the lesion. However, the dynamics of hormone synthesis and transport were more clearly demonstrated in the 1960s by the experiments of Sachs, Sloper and their respective colleagues, using radio-tracer and autoradiography techniques (see Chapter 2 Section 4(v)) to follow the distribution of injected [^{35}S] cysteine. The amino acid cysteine is a major constituent of the neurohypophyseal hormones (see Figure 3.1) and the neurophysins. When [^{35}S]cysteine was injected into rats, either into the cerebral ventricles or the hypothalamus, it rapidly accumulated in the SON and PVN and was later shown to be converted into labelled vasopressin or oxytocin, as well as neurophysin. After 1 to 2 hours, radioactive material appeared in the infundibular process and later still appeared in the posterior lobe at a time when levels in the hypothalamus were depleted. On the basis of such experiments, rapid axonal transport of hormone was determined (estimated to be approximately 2–3 mm/h). Furthermore, in rats given a hypertonic salt solution to stimulate vasopressin secretion, a faster accumulation and transport of labelled amino acid occurred. Similar incorporation of other radioactively labelled amino acids has confirmed the rapid synthesis and axonal transport of oxytocin and vasopressin. That the hormones are packaged and transported in secretory granules was also confirmed by a similar method, since injected labelled phosphorus (^{32}P) which is largely incorporated into granule membranes, showed an identical rate of axonal transport.

Axonal transport within neurones in general is a major topic of research interest and the reader is referred to [13] for further information. It is an active process involving specialised microtubules which appear to conduct substances within granules or vesicles at different rates. Transport of substances towards the cell body (retrograde flow) is also well documented. With respect to axonal transport of the neurohypophyseal hormones which have been thoroughly studied, it is possible that vasopressin and oxytocin are transported at different rates. However, this is difficult to distinguish from differences in the rate of processing of hormone from precursor since other evidence suggests that vasopressin is processed faster than oxytocin. It is however of interest, that contrary to what one might expect, the rate of axonal transport of the neurohypophyseal hormones is usually unaltered by changes in the rate of secretion (see Chapter 3 Section 4). For further reading on the biosynthesis of the neurohypophyseal hormones see [8, 10, 12].

3:4 Neurophysins

(i) *Hormone Carrier Proteins*

It was known from the earliest studies of the neurohypophyseal system that the transported secretory granules contained large amounts of protein in addition to the smaller peptide hormones. Many studies of the neurohypophyseal hormones, some of which were outlined in the previous section, indicated that a significant proportion of this protein in the granules was specifically associated with the synthesis and transport of the hormones. These proteins have been called neurophysins. They are linear cysteine-rich molecules consisting of approx-

Figure 3.2: Proposed Mechanism for the Transport and Release (Exocytosis) of Neurohypophyseal Hormones. For explanation see sections 3:4 and 3:6. NPH(P) — neurophysin hormone complex (precursor); NP — neurophysin; H — hormone; K — protein kinase; M — myosin-like protein; A — actin-like protein.

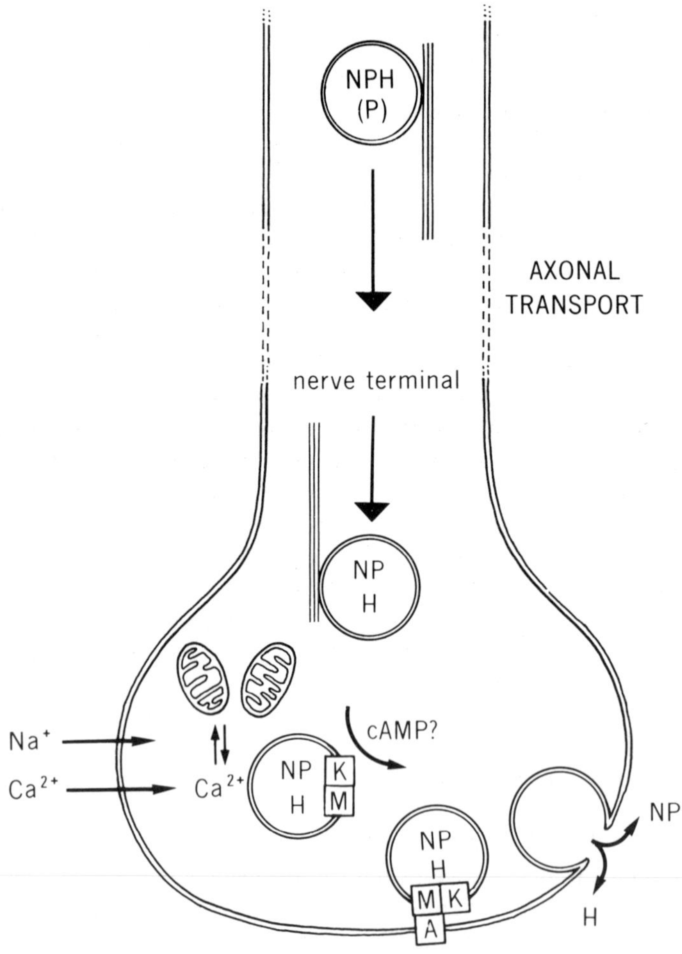

imately 95 amino acid residues, but their precise structure has only been fully characterised in a limited number of species. Most animal species studied appear to possess two types of neurophysin specifically associated with the two neurohypophyseal hormones, although only one and sometimes three have been reported in some species. Much evidence indicates that neurophysins function as 'carrier proteins' for oxytocin and vasopressin, and thereby prevent the loss of the readily diffusible peptide hormones prior to their secretion. Evidence also suggests that each neurophysin–hormone complex is processed in specific secretory granules from a common precursor molecule which is then split into neurophysin and hormone during axonal transport (see Figure 3.2). The radioactive tracer studies with [^{35}S] cysteine, described above, are consistent with a common molecular processing of neurophysin and hormones. For example it has been shown that [^{35}S] cysteine, injected adjacent to the SON in rats, was incorporated into a single protein of a molecular weight of about 20,000. After 2 hours, to enable transportation of material down to the pituitary, a further labelled protein of molecular weight about 12,000 was found in the neurohypophysis and also in the median eminence. After 24 hours, only the smaller labelled protein was present in the neurohypophysis and, in addition, labelled oxytocin and vasopressin, as well as several other smaller peptides were present. These results generally support the view that synthesis of a large precursor protein occurs and this is rapidly cleaved in the granules during axonal transport to give a separate neurophysin protein and hormone in the pituitary (see Figure 3.2). The presence of enzymes in the granules seems to be responsible for this process, but these remain to be fully characterised and identified (see [1] and [11] for further details).

(ii) *Neurophysin Localisation*

The development of specific antibodies to the neurophysins has enabled precise localisation of neurophysin in the brain by immunohistological techniques. In the SON most of the cells contain neurophysin whereas only about 75% of the large cells located in the PVN contain these proteins. Neurophysin is also found all along the axons of the supraoptico-hypophyseal tract and, although the majority of these fibres terminate in the posterior lobe of the pituitary, a significant number of neurophysin-containing axons also terminate in the zona externa of the median eminence in close proximity to the hypophyseal portal capillaries. Using the technique of pituitary stalk interruption or constriction, immunofluorescent labelled neurophysin was found to accumulate only on the cell body side of the lesion and the axonal flow of the neurophysin was found to be similar to that reported for the hormones (i.e. 2–3mm/h). In addition, neurophysin-specific antibodies have been used to reveal two separate groups of cells in the SON and PVN which presumably relate to separate cells which synthesise vasopressin and oxytocin. Further evidence for neurophysin-hormone specificity is the demonstration that in the mutant strain of Brattleboro' rats, which have congenital diabetes insipidus and lack vasopressin in their neurohypophyses, also lack the vasopressin-associated neurophysin.

(iii) *Biochemical Studies*

Neurophysins isolated from various mammalian species including rat, pig and bull have clearly identified hormone specific neurophysins. Generally, the biochemically isolated neurophysins will non-specifically bind and form complexes with either vasopressin or oxytocin and in similar ratios to the natural isolated complex. Because of the lack of specificity in hormone binding, neurophysins have been classified into two types by the following criteria: (a) amino acid composition, (b) mobility of separation by electrophoresis and (c) by their release from the neurohypophysis in response to pharmacological stimuli. Using neurophysin-specific radioimmunoassays it has been shown in humans and monkeys that one particular neurophysin is released with oxytocin in response to oestrogen whilst the other is released together with vasopressin in response to haemorrhage or nicotine. Clearly, neurophysins are co-stored with the posterior pituitary hormones and released along with the hormone following induced secretion. For further reading see [1, 11, 12].

3:5 Storage of Posterior Pituitary Hormones

As described in the previous sections, the posterior pituitary hormones and their associated neurophysin proteins are synthesised and transported at high rates towards the nerve terminals in the neurohypophysis. However, both the synthesis and the rate of transport of hormone are not directly or immediately altered by changes in secretion as one might suppose, and furthermore only a small proportion of hormone content of the neurohypophysis can be released at any one time following maximum stimulation. The reasons for this have been demonstrated relatively recently and relate to the processing of hormone-containing storage granules in the neurohypophysis.

As has been demonstrated for many neurotransmitter and neurosecretory systems, it is the most recently synthesised hormone or transmitter or the most recently arrived storage granules which are released first. This is most simply demonstrated in a system in which a radioactive label (e.g. [^{35}S] cysteine for the posterior pituitary hormones), is incorporated into the releasable material immediately prior to stimulation of release. Invariably, the specific radioactivity of released hormone (i.e. the proportion which is labelled) is higher than that of unreleased labelled hormone remaining in the terminals.

Figure 3.3: Proposed Movement of Neurosecretory Granules in the Neurohypophyseal Nerve. For explanation see section 3:5.

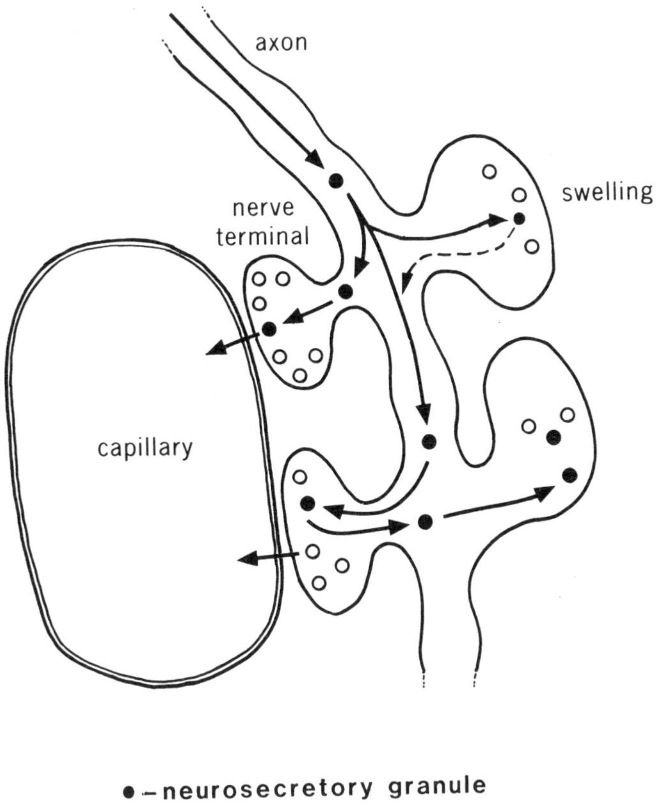

Detailed electron micrographic studies of the rat neurohypophysis have revealed complex terminals with several neurosecretory sites and many associated axonal vacuoles or swellings. The movement of radioactively labelled granules in these terminals has been studied (Figure 3.3). Newly arrived granules congregate at neurosecretory sites in preparation for release into neighbouring blood capillaries, but those which are not soon released move to the non-releasable regions (swellings) where the action of lysosome enzymes break down the granules and their contents. It has been proposed that in a neurosecretory system where secretion is discontinuous, as in the neurohypophysis, that the maintenance of readily releasable hormone may be finely regulated by changes in rates of disposal of stored hormone. This, albeit wasteful mechanism of control, is clearly important with large neurosecretory neurones where the site of secretion is far removed from that of hormone synthesis. Whether this process is characteristic of other neurosecretory neurones such as those of the tuberoinfundibular system remains to be established by further research. For further reading see [7].

3:6 Neurosecretion of Neurohypophyseal Hormones

The process by which nerve cells secrete a peptide hormone into the circulation has been termed stimulus-secretion coupling. It can be defined as a transduction of a neural signal (depolarisation of the nerve endings) to a hormonal signal (secretion of the hormone). Studies on the release of biologically active peptides from neurones have been most extensively carried out on the neurohypophyseal system, partly because isolated neurohypophyses, which contain the axons and terminals of the hypothalamic–neurohypophyseal tract, lend themselves readily to *in vitro* studies in which various experimental parameters can be simply controlled. Although this section deals specifically with stimulus-secretion coupling in the neurohypophysis, it is generally agreed that similar mechanisms of hormone release exist in the hypothalamic neuroendocrine cells that control adenohypophyseal secretion via the hypophyseal portal blood, i.e. those of the tuberoinfundibular system.

(i) *The Stimulus or Trigger for Secretion*

Electrical activation of the neurosecretory cell is the stimulus for neurosecretion, and this electrical activity is determined by the sum of excitatory and inhibitory synaptic inputs which impinge on to the cell body. The electrical properties of neurosecretory cells appear to be similar to all other neurones in that depolarisation of the cell body, as a result of neural inputs transmitted to the cells, leads to the generation of action potentials which travel along the nerve axon to the terminal. Due to the unequal distribution of ions (chiefly potassium and sodium) between the extra- and intracellular compartments, a potential exists across the membrane of nerve cells, the inside being negative with respect to the outside. In neurosecretory cells this 'resting' membrane potential is in the order of 60mV. The action of neurotransmitters, released from the synaptic terminals of afferent fibres, is to alter the membrane permeability of the postsynaptic neurone. When the permeability (conductance) to sodium ions is increased, they enter the cell, down both a concentration and electrical gradient, and the inside of the neurone becomes less negative (depolarisation).

If the depolarisation reaches a critical 'threshold' level, an action potential (wave of depolarisation) is generated in the axon of the nerve cell. The conduction velocity of the action potentials in neurosecretory cells is in the order of 1 m/s which is relatively slow when compared with conduction velocities of over 100 m/s in large motor axons. The resting membrane potential is restored when the permeability to sodium becomes low again and potassium ions pass out of the axon. The arrival of the action potential at the nerve terminal results in the depolarisation of the terminal, which leads to the inward flux of sodium and calcium ions thus triggering the process of stimulus secretion coupling (see [7, 10] for further information).

(ii) *Mechanism of Hormone Discharge – Exocytosis*

Light and electron microscopic techniques have demonstrated the presence of both large (2000 Å diameter) and small (ca.400 Å) secretory vesicles within the nerve terminals of neurohypophyseal neurones. In addition the small vesicles were shown either to be coated or smooth, the latter resembling synaptic vesicles. The large vesicles or granules were shown to contain the posterior pituitary hormones, as discussed previously, but the presence of small vesicles led to considerable speculation. One consideration, later disproved, was that the vesicles contain a neurotransmitter which regulates secretion of the hormones. However, stimulation of the neurohypophysis was shown to result in a decrease in the number of granules and an increase in the number of both coated and smooth small vesicles. This evidence, together with other microscopic observations and subcellular fractionation techniques, demonstrated that the neurohypophyseal hormones, together with their neurophysins, were indeed contained within the granules whereas the small vesicles were formed from the membranes of large vesicles following secretion of their hormone content. It is now well established that neurohypophyseal hormones are discharged by a process of exocytosis, although alternative mechanisms of release have been proposed. Exocytosis involves fusion of the granular and cellular membranes, consequent rupturing of the membranes and release of hormones together with total granular content. The granule membrane is then retrieved from the terminal surface by a process of pinocytosis (effectively the reverse of exocytosis) with the formation of coated microvesicles. The coating is eventually shed with the formation of smooth synaptic-like vesicles. Usually these are subsequently incorporated into lysosomes for disposal. A scheme depicting exocytosis in nerve terminals of the magnocellular cells is shown in Figure 3.2.

The evidence that the neurohypophyseal hormones are secreted by exocytosis is firstly ultrastructural. The process of exocytosis is difficult to observe by conventional electron microscopy, but freeze fracture electron microscopy (see Chapter 2 Section 2(iii)) of tissue obtained from the maximally stimulated isolated neurohypophysis or on isolated neurohypophyseal terminals (synaptosomes – see Chapter 2 Section 5(i)) clearly provides evidence that neurosecretion involves the process of exocytosis. Fusions of secretory vesicles both with the cell membrane and with each other and characteristic membrane alterations such as 'necks' and 'holes', which are associated with exocytosis, have been observed in neurohypophyseal tissue. However, exocytotic profiles are not frequently found but this is not so surprising when one considers that even during extreme stimulation it is estimated that, in the rat, each terminal releases only 1 or 2 granules per minute.

Biochemical evidence for exocytosis comes from the careful analysis of substances which are released during secretion and those which are not. As indicated previously, posterior pituitary granules contain neurophysins associated with the hormones. It has been shown that hormone and neurophysin are always released in constant proportions during secretion whereas the release of cytoplasmic substances does not occur. The findings from both ultrastructural and release studies are consistent with the concept that exocytosis involves the release of the total granular content during secretion.

(iii) *Role of Calcium Ions (Ca^{2+}) in Neurosecretion*

The intracellular resting level of free Ca^{2+} in the neurohypophysis is probably in the same range as that reported for other neurones (50–150nM). Overwhelming evidence exists that a small rise in intracellular free calcium is an essential prerequisite for the release of peptide hormones and neurotransmitters in general. Depolarisation of the nerve endings, either electrically or by subjecting the tissue to a relatively high concentration of potassium ions, causes release of the peptides and this effect is blocked when Ca^{2+} is absent from the external medium. However, diffusion of Ca^{2+} across the cell membrane, as a result of depolarisation, does not appear to be the only way in which the intracellular concentration of free Ca^{2+} is raised and various amplification mechanisms, which are triggered by Ca^{2+} moving across the cell membrane, have been proposed. These include (a) release of Ca^{2+} from binding sites in the plasma membrane, (b) release of Ca^{2+} from the endoplasmic reticulum and (c) inhibition of mitochondrial uptake of Ca^{2+} due to the rise of intracellular sodium ions. It is as yet unknown how large the free Ca^{2+} concentration may become in the region of exocytosis but the rise in intracellular Ca^{2+} concentration may be very localised within the nerve terminal.

There is a wealth of experimental studies which have been directed toward establishing how the increased Ca^{2+} concentration leads to neurosecretion and as a result various mechanisms have been suggested which include:

(a) Calcium reduces the electrostatic barrier between granule and cell membranes.
(b) Calcium ions reduce the cytoplasmic viscosity to assist granule and cell membrane fusion.
(c) Calcium activates a hypothetical enzyme which triggers fusion of granule and cell membrane.
(d) Calcium ions accelerate movement of granules along microtubules.

Currently the most favoured hypothesis is the involvement of microtubules in the process of neurosecretion. Evidence for microtubules both within nerve endings and nerve axons has been presented and it has been proposed that microtubules or microfilaments may propel granules and cell membrane into fusion by some contractile process.

From striated muscle it is known that Ca^{2+} binding proteins can serve as regulators of the contractile mechanisms and recently a number of analogies have been pointed out between stimulus-contraction coupling and stimulus-secretion coupling. In muscle it is well established that depolarisation of the muscle membrane results in the release of free Ca^{2+} from the sarcoplasmic reticulum. This Ca^{2+} then combines with a protein (troponin–tropomyosin) complex on the actin filaments; such binding removes an inhibition and allows the actin filament to interact with a charged myosin–ATP complex. A bridge is consequently formed, ATP is broken down to ADP and ultimately the myosin filament 'slides' along the actin filament thus bringing about a contraction. Parallel mechanisms have been proposed for the movement of the secretory granules to the nerve terminal along the microtubules and for the fusion of the granule membranes with the terminal membranes.

Actomyosin complexes, similar to muscle have been isolated from secretory tissue and the release of hormones has been shown to involve the breakdown of ATP by a granule associated ATPase, indicating the involvement of an energy-consuming process. In more recent studies a protein showing similar properties to myosin has been isolated from membranes of neurosecretory granules, whereas an actin-like protein appears to be present in the nerve terminal membrane; a troponin–tropomyosin complex is present in the axoplasm. Thus, analogous to muscle, Ca^{2+} in the nerve terminal may combine with the troponin–tropomyosin binding protein thereby allowing the granule-associated myosin to interact with the actin in the nerve terminal membrane resulting in membrane fusion and exocytosis (Figure 3.2).

Apart from the involvement of Ca^{2+} in neurosecretion, there is also evidence that the cyclic nucleotide, cyclic AMP, (an intracellular mediator of many hormonal actions, see Chapter 1 Section 10) may also be involved. Experiments have shown that neurosecretory granules contain cAMP, and recently a cAMP-stimulated protein kinase has been isolated from secretory granules. This kinase is capable of phosphorylating membrane-associated proteins and it has been postulated that there may be some interaction with cAMP, Ca^{2+} and the phosphorylation of membrane-linked myosin. However, as yet no definitive role of the cyclic nucleotide in secretory processes has been established although experimentally cAMP can stimulate secretion in a wide spectrum of tissues.

There is no doubt that Ca^{2+} is the trigger for neurosecretion, but although the rise in intracellular Ca^{2+} may be small and may remain localised, there must also be mechanisms which normalise the concentration following stimulation. There is an ATP-dependent Ca^{2+} uptake by the mitochondria and in the bovine neurohypophysis a similar uptake by the endoplasmic reticulum has been observed. Passive Ca^{2+} may also occur in the secretory granules themselves. Such intracellular sequestration would enable a rapid reduction of free Ca^{2+}, but eventually the excess Ca^{2+} must be transported back out of the cell. This process is not well clarified although both an ATP-dependent Ca^{2+} pump and a sodium-transport-dependent calcium transport have been proposed.

For further reading on the secretion of the neurohypophyseal hormones see [4, 7, 10, 14].

3:7 Physiological Control of Neurohypophyseal Hormone Secretion

(i) *Oxytocin Secretion*

The release of oxytocin is generally triggered by stimuli associated with reproductive functions (see Figure 3.4). Stimulation of the female genitalia will elicit the release of oxytocin although the significance of this response remains obscure. At parturition oxytocin plays an important role in stimulating contractions of the uterus, although it is now known that oxytocin is not responsible for initiating labour but plays a supportive role. One current line of thought is that the high levels of oestrogen may increase the number and/or affinity of oxytocin receptors in the uterus and that the interaction of this hormone with its receptors induces prostaglandin release; certain prostaglandins are potent stimulators of uterine smooth muscles. The neural stimulus for oxytocin secretion during labour arises from stretch receptors in the uterus and possibly in the mucosal wall of the vagina. These receptors initiate action potentials which are relayed via a polysynaptic pathway to the oxytocin neurosecretory cells in the hypothalamus.

The most widely investigated stimulus for oxytocin secretion is that of suckling as occurs during lactation. As early as 1910 Ott and Scott had demonstrated that neurohypophyseal extracts had milk-ejecting activity and by 1940 the modern concept of the milk-ejection reflex had been formulated. In most species of mammals so far studied, (with the exception of some domestic ruminants) the secreted milk, which is stored in the alveolar lumina and small ducts of the mammary gland, can only become available to the suckling pup through a neuroendocrine reflex – the milk-ejection reflex. This reflex has a neural afferent path and a hormonal efferent path. Suckling stimulates sensory receptors in the nipple and the sensory afferents, which ascend ipsilaterally in the spinal cord and pass through the midbrain and to the hypothalamus, relay neural information to the oxytocin neurones. Their stimulation results in secretion of the hormone which acts on

Figure 3.4: The Control of Oxytocin Secretion. ACh – acetylcholine; NA – noradrenaline; NP – neurophysin.

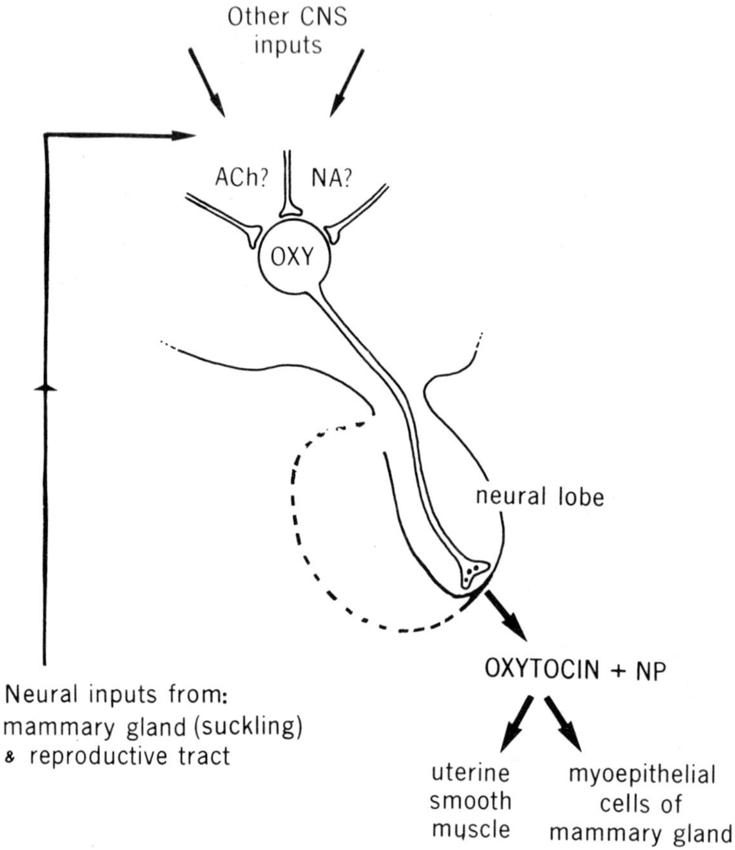

the myoepithelial cells in the mammary glands. Contraction of these cells results in milk ejection or 'let-down'. The release of oxytocin from the neurohypophysis occurs in a quantal fashion and the neural activity associated with the spurts of hormone secretion has been fully characterised by elegant experiments on the suckling rat (see Chapter 3 Section 8).

(ii) *Vasopressin Secretion*

It has long been known that the anti-diuretic hormone, arginine vasopressin (see Figure 3.1) plays an important role in regulating water reabsorption in the kidney (see section 3.9). Until recently the relationship between vasopressin secretion and water balance was poorly defined due to the difficulties of measuring, by bioassay, the low concentrations of vasopressin normally found in blood and the body fluids. This difficulty has been overcome following the development

Figure 3.5: The Control of Vasopressin (VP) Secretion. ACh – acetylcholine; DA – dopamine; NA – noradrenaline; NP – neurophysin; VP – vasopressin.

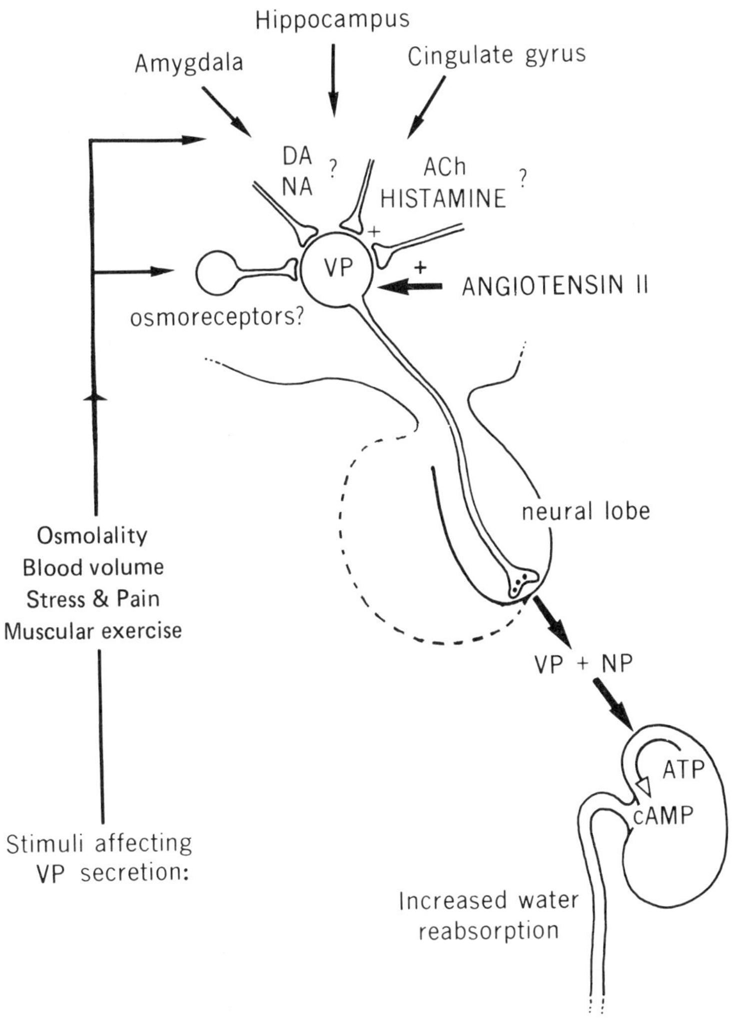

of sensitive and specific radioimmunoassays (see Chapter 2 Section 4(iii)) coupled with appropriate extraction procedures, and it is now possible to characterise more precisely the relationship between vasopressin secretion and states of water balance.

The primary physiological parameters in the control of vasopressin release are the volume and osmotic composition of the extracellular fluids (see Figure 3.5). A reduction of body fluids, as occurs during haemorrhage, stimulates vasopressin release with a consequent increase in water reabsorption from the collecting ducts of the kidneys and restoration of fluid volume. On the other hand an increase in plasma osmolality and its principal determinant, plasma sodium, also stimulates vasopressin secretion so that the resulting increased water retention by the kidney serves to dilute the extracellular fluid and reduce the plasma osmolality. A reduction of blood volume is a far more potent stimulus for vasopressin release, in terms of the plasma concentrations that can be achieved, than is an increase in the osmotic pressure of extracellular fluids although the relative contributions between osmotic and volume changes that stimulate vasopressin release are not clearly defined.

The functional properties of this system in the 'normal' physiological state can be appreciated most easily by examining the relationship between the four major variables: plasma osmolality, plasma vasopressin, urine osmolality and thirst. In healthy adults plasma osmolality correlates closely with plasma vasopressin. Below about 280 mOSM/kg vasopressin secretion is suppressed and the low circulating levels (less than 1 pg/ml) enables maximum diuresis (i.e. urine formation about 10 ml/min). Above this value of osmolality the secretion of vasopressin rises steeply and proportionately with osmolality, reaching about 5 pg/ml plasma at around 295 mOSM/kg. This plasma concentration is sufficient to produce maximum antidiuresis (i.e. maximum solute-free water reabsorption from the urine). At the same level of plasma osmolality healthy adults have a conscious desire to drink and such a thirst mechanism compensates dehydration and further rises in osmolality. Hence the threshold for vasopressin secretion and osmotic stimuli for thirst perception normally determine the lower and upper limits respectively to which plasma osmolality is allowed to rise and fall.

It is now known that changes in blood osmolality are detected by osmoreceptors. The term osmoreceptor was first adopted by Verney in 1947 following his observations that an increase in the osmolarity of brain blood produced by an intracarotid injection of hypertonic saline elicited vasopressin release from the neurohypophysis. He suggested that specialised brain cells were sensitive to changes in osmotic pressure and this was later confirmed by electrophysiological experiments (see section 3:8). However, not all solutes appear to be able to stimulate osmoreceptors despite increasing osmolality. Sodium and mannitol, for example, are potent stimulators although glucose can actually inhibit vasopressin secretion (e.g. as in diabetes mellitus) despite rising plasma osmolality. Such observations, coupled with the fact that sodium and its associated anions contribute over 95 per cent of the osmotically active constituents of the plasma, have led to the concept that cells in the hypothalamus are sensitive to Na^+ concentrations rather than osmolality. However, the evidence for and against sodium receptors versus osmoreceptors is controversial and remains to be established.

Changes in the volume of body fluids which can also alter the rate of vasopressin secretion are detected by stretch receptors in the cardiovascular system. In 1938 Rydin and Verney showed that removal of a small amount of blood from a conscious dog resulted in an inhibition of water diuresis although there was only a transient fall in the mean arterial blood pressure. This effect of haemorrhage was subsequently confirmed by other experiments and it was shown that the amount of vasopressin secreted is proportional to the decrease in blood volume in the range of 10–50 per cent, a 9–10 per cent reduction being the threshold for secretion. Two groups of receptors have now been identified as factors in this neuroendocrine reflex. Firstly, stretch (volume) receptors in the walls of the left atrium and secondly the arterial baroreceptors in the aortic arch and carotid sinuses. A fall in blood volume results in a decrease in the number of action potentials generated in these receptors which consequently results in a decreased tonic inhibition of the neurosecretory cells normally imposed by volume and pressure receptors – vasopressin secretion is thus increased. Conversely, a rise in extracellular fluid volume increases the inhibitory effect on the neurosecretory cells with a suppression of vasopressin release. Recently the existence of other volume receptors in the abdomen has been proposed although they have not yet been identified.

Cutaneous and somatic efferent nerves also relay stimuli to the vasopressin neurosecretory cells as non-specific factors such as pain and muscular exercise release vasopressin. In addition vasopressin secretion can be stimulated by emotional stress and this response may involve the limbic system. In certain situations stimuli that release vasopressin can also release oxytocin and vice versa and both hormones possess intrinsic activity of the other hormone due to their structural homology (see Figure 3.1). For further reading on the factors regulating the secretion of vasopressin see [4, 5, 6].

3:8 Electrophysiological Studies of Neurosecretory Cells

(i) *Magnocellular Neurones of the Neurohypophyseal System*

In common with the anatomical and biochemical studies described above, the magnocellular neurones of the mammalian hypothalamus have a number of features which make them suitable for electrophysiological investigation. The neurosecretory cells in the supraoptic nuclei (SON) and paraventricular nuclei (PVN) are densely packed, their diameter is large allowing for intracellular and extracellular recording, their axons run in a well-defined bundle and they terminate in a well-defined structure, the posterior pituitary. In addition, specific and reliable bioassays and radioimmunoassays are now available for the measurement of circulating hormone concentrations and thus blood samples, obtained during electrical recording, enable correlations between electrical activity and hormone release. In some instances hormone secretion can be assessed on-line during electrophysiological experiments and, in the case of oxytocin, this has led to quantitation of hormone secretion associated with each action potential generated in the neurosecretory cells.

The identification of neurosecretory cells by antidromic stimulation techniques (see Chapter 2 section 3(iii)) proved to be a turning point for electrophysiological studies on magnocellular neurones. Both extracellular and intracellular studies have indicated that these cells display orthodromic, antidromic and postsynaptic responses which are indistinguishable from other CNS neurones and the neural events associated with a variety of stimuli known to involve the release of oxytocin and/or vasopressin have been extensively investigated on identified neurosecretory neurones. Electrophysiological studies have also suggested that these neurosecretory cells have axon collaterals (see Chapter 2 Section 3(iii)). The collateral may either course back to the cell body making direct or indirect (via an interneurone) synaptic contact or make contact with adjacent neurosecretory cells. The significance of such collaterals is that they may play a role in the pulsatile pattern of hormone release by causing recurrent inhibition of the parent neurone or they may assist in the simultaneous activation of adjacent neurones, thus ensuring sufficient secretion of a hormone at one point in time.

(ii) *Oxytocinergic Neurones*

In 1966 it was first demonstrated that a gentle intermittent suction applied to the nipples of female rats increased the spontaneous firing rate of some 80 per cent of the PVN neurones tested in the hypothalamus. The change in unit firing was accompanied by a rise in intramammary pressure, indicative of oxytocin secretion, and this was the first demonstration of a definite correlation between electrical activity of hypothalamic neurones and the secretion of a pituitary hormone.

Subsequently, Wakerley and Lincoln, during the early 1970s, analysed the neural activity associated with the suckling stimulus and oxytocin release. Their elegant experiments are worthy of a brief description. They recorded from antidromically identified neurosecretory cells in the PVN of anaesthetised lactating rats during suckling of the pups. Maternal milk ejection, which is indicated both by a 'stretch' response from the young and a rise of intramammary pressure, occurred every 4–8 min even though suckling was continuous. Approximately 18 seconds before the rise of intramammary pressure the paraventricular units showed a sudden increase of firing rate from an average of 2 spikes/second to 30–50 spikes/second. This burst of activity lasted 2–4 seconds then decayed exponentially to a period of after inhibition lasting between 7 and 56 seconds (see Figure 3.6). The fact that hormone release is dependent on the frequency of firing (30–50Hz) has been well supported by electrical stimulation studies of the neurohypophysis which have shown that stimuli in excess of 30 Hz are required to produce any appreciable hormone secretion. In an attempt to quantify neural activity and hormone release, it was estimated that the total amount of oxytocin released prior to each milk ejection was 1/2 mU which represented 0.2 per cent of the total quantity of oxytocin in the neurohypophysis. This indicated that an average 1 in 1000 secretory granules were released at each milk ejection and that during the high burst of activity one spike released 3×10^{-5} g of oxytocin. These experiments represent the first quantitative approach to the relationship between neural activity and neurosecretion. Other stimuli, such as vaginal distension, will also cause oxytocin secretion in suckling rats although it is a less effective stimulus since it is accompanied by a less marked enhancement of neuronal activity and a smaller rise in intramammary pressure.

Figure 3.6: Electrophysiological Studies of the Milk-ejection Reflex (by courtesy of D. W. Lincoln). For explanation see Section 3:8.

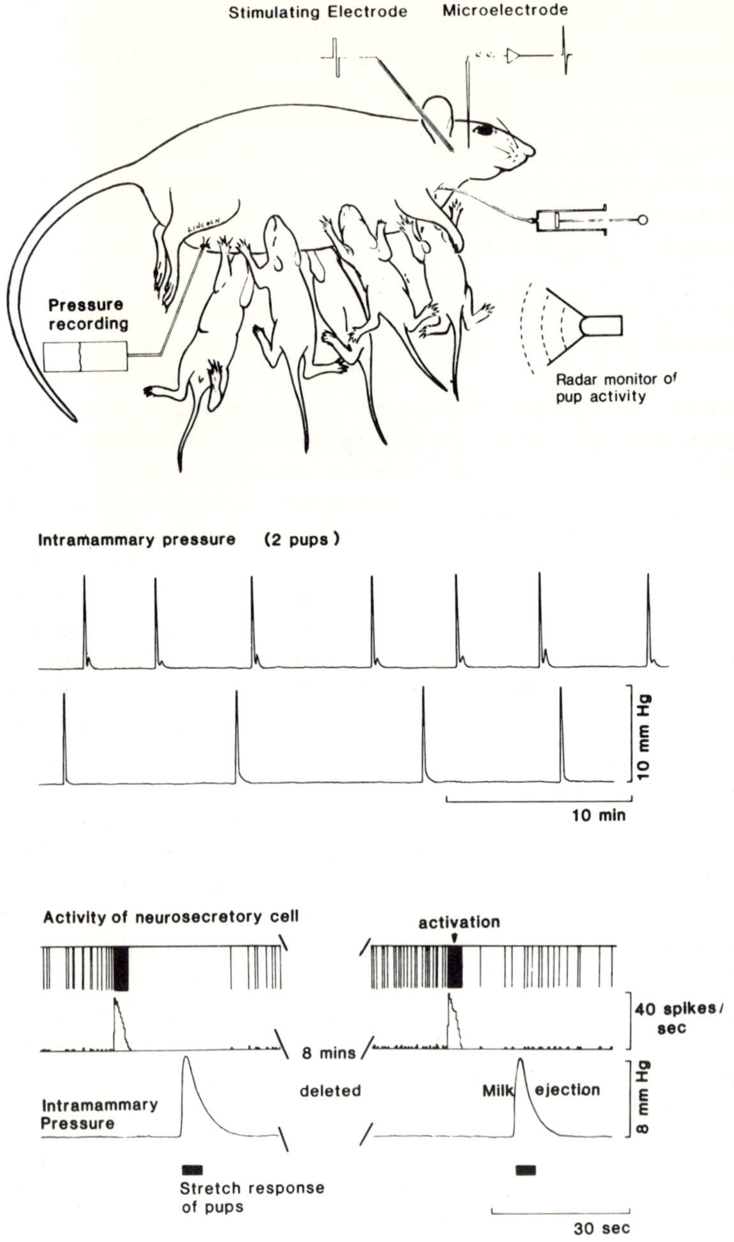

(iii) *Vasopressin Neurones*

In 1959, Cross and Green showed that neurones in the SON displayed an increased rate of firing after injection of hypertonic saline. Furthermore, neurones responding to osmotic stimuli were highly specific and generally they were unaffected by suckling tactile, auditory, or visual stimuli. It has since been demonstrated that approximately 25 per cent of antidromically identified supraoptic and paraventricular neurones display phasic bursts of activity and that such cells are not activated by suckling but only by stimuli which are known to evoke vasopressin release, such as osmotic infusions, dehydration, occlusion of the carotid artery or haemorrhage. Based on these observations it has been suggested that the neurosecretory neurones which show phasic responses are vasopressin-containing cells.

Unlike oxytocin cells which show a distinctive and characteristic response to suckling, vasopressin cells appear to show a variety of physiological responses to stimuli that evoke its release. For example, in the rat hypertonic saline has been shown to enhance the overall electrical activity although the periodicity of phasic discharge is retained. A continuous pattern of firing has been reported, however, following withdrawal of blood with the phasic pattern returning only after the blood volume is restored to normal. Conversely experiments on the anaesthetised monkey have suggested that phasic discharges are just one functional state of activity in these neurones. Under conditions of progressive dehydration, the pattern of firing changes from irregular low-frequency discharges to phasic discharges to high-frequency continuous discharges in concert with increasing osmolarity. One intriguing question is the nature of the osmosensitive neurone itself. Electrophysiological experiments have suggested that the osmoreceptors are not the neurosecretory cells but interneurones which lie in the perinuclear zone of the SON. Such cells respond specifically to osmotic stimuli but could not be excited antidromically by stimulation of the neurohypophysis. It has been proposed that these cells relay 'osmotic information' to the adjacent neurosecretory cells, although other data imply that the neurosecretory cells may also be osmosensitive. For further details of electrophysiological approaches to the neurohypophyseal hormones see [2, 15].

3:9 Neurotransmitters and Neuropeptides

(i) *Neurotransmitter Regulation of the Magnocellular Neurones* (see Figures 3.4 and 3.5).

Neural inputs from the mammary gland, the female reproductive tract and volume receptors have been described in Sections 3:7 and 3:8. In addition, electrical stimulation of several hypothalamic areas and a number of extrahypothalamic regions, which form part of the limbic system (including the amygdala, hippocampus and cingulate gyrus) induce the release of the neurohypophyseal hormones and alter the firing frequency of neurosecretory cells (see Figure 3.5). This is consistent with the anatomical evidence for connections between the magnocellular nuclei and the cortex, amygdala and septum. However, widespread destruction of these brain areas does not inhibit the milk-ejection reflex suggesting that these areas may be involved with emotional or stress-induced mechanisms of hormone release.

There is good evidence that some of the afferent connections with the neurosecretory cells are cholinergic in nature i.e. are mediated by the neurotransmitter acetylcholine (ACh). Intracarotid injections of ACh stimulate the secretion of both vasopressin and oxytocin and anticholinergic drugs can inhibit the milk-ejection reflex. Iontophoretic studies have confirmed that neurosecretory cells are excited by local application of ACh and that both types of cholinergic receptors, muscarinic and nicotinic, are involved in these responses.

The paraventricular nucleus contains numerous noradrenergic and adrenergic nerve endings (i.e. contain the neurotransmitters noradrenaline and adrenaline) and thus it is not surprising that cells in this area are sensitive to the microiontophoretic application of noradrenaline (see Chapter 2 Section 3(iii)). Generally noradrenaline inhibits the firing of neurosecretory cells in the PVN and SON and these responses may be blocked by noradrenoreceptor antagonist drugs. Other common neurotransmitters in the hypothalamus, such as dopamine and 5-hydroxytryptamine (5-HT) either enhance or depress the activity of magnocellular cells but since there are relatively few dopamine or 5-HT nerve terminals in either the PVN or SON these effects may not be significant. Histamine is also thought to function as a neurotransmitter in the hypothalamus and other brain regions. When histamine is injected into the third ventricle close to the hypothalamus it promotes antidiuresis and when directly applied by microiontophoresis on to either supraoptic and other hypothalamic neurones excitation of the cells has been observed. This is in contrast with other areas of the brain where iontophoresis of histamine on to single cells characteristically depresses neuronal firing. The hypothalamus contains the highest concentration of histamine in the brain, and its possible role in the release of vasopressin secretion has been demonstrated.

Interestingly, iontophoretic studies have also shown that the neurosecretory cells are sensitive to their own hormones. Vasopressin inhibits the firing rate of most responsive neuroendocrine cells but only at relatively high microiontophoretic currents (see Chapter 2 Section 3(iii)). On the other hand, oxytocin results in a dramatic increase in the firing rate of oxytocin-secreting neurones. Although it was initially thought that these responses may be associated with the axon collaterals of neurosecretory cells there is no evidence as yet to support this concept and any neurotransmitters involved in the recurrent inhibition remain to be identified.

Finally mention should be made of centrally acting neuropeptides. Firstly angiotensin II, a peptide produced in blood which is important in the regulation of water balance and blood volume and is also found in brain. Iontophoretic application of angiotensisn II to supraoptic neurosecretory cells increases their firing rate and has also been shown to stimulate vasopressin secretion. In addition, intracranial injections of angiotensin II into the anterior hypothalamus, septum or preoptic area or into the cerebral ventricles immediately causes the animal to drink (see Chapter 11 Section 9). A further neuropeptide reported to centrally-stimulate vasopressin release is β-endorphin (see Table 1.4 and Chapter 4 Section 5) but the importance of this peptide in regulating antidiuretic activity and vasopressin release requires confirmation.

(ii) *Neurotransmitters and Neuropeptides in the Posterior Pituitary*

Cholinergic neurones have been demonstrated in the neurohypophysis and for some time there was a debate whether the acetylcholine (ACh) was present in the small vesicles of the neurosecretory nerve-endings and could be implicated in regulation of hormone secretion. However, present evidence is in favour of separate ACh neurones with a vasomotor function, since, for example, ACh causes vasodilation of capillaries in the neurohypophysis and thereby indirectly facilitates peptide hormone secretion.

There have been a number of reports that low levels of neuropeptides other than the neurohypophyseal hormones are located in the neurohypophysis (e.g. Met- and Leu-enkephalin, dynorphins, cholecystokinin (CCK-8), neurotensin, somatostatin, angiotensin II and corticotrophin releasing factor). At the present time, little is known regarding the function of such peptides and whether they regulate hormone secretion or are secreted into the systemic circulation along with the neurohypophyseal hormones. However, in contrast with β-endorphin (see above) the natural opioid peptides, Met- and Leu-enkephalin, inhibit both oxytocin and vasopressin release. Their presence in the neurohypophysis has been determined by radioimmunoassay and immunohistochemical techniques, whilst binding studies indicate the presence of receptors for these opioid peptides in the gland. A very recent cytochemical study has suggested that the enkephalins co-exist with the posterior pituitary hormones in the same nerve terminals – invariably Met-enkephalin was found in association with oxytocin and Leu-enkephalin with vasopressin. Whether opioid peptides are truly found in the hormone–neurophysin containing secretory granules or separate granules and how they may act within the nerve terminal to inhibit neurosecretion remains to be determined. Nevertheless, in the light of growing evidence that neuropeptides may act as neuromodulators or neurotransmitters (see Chapter 10 Section 3), the finding that opioid-peptides co-exist with hormones in the terminals of neurosecretory cells may provide a useful system to study cellular mechanisms of the peptide neuromodulators. For further reading on the neural control of the posterior pituitary see [6].

3:10 Actions of Neurohypophyseal Hormones

This section summarises the cellular actions of the posterior pituitary hormones on their target tissues but the description will be brief since the actions of hormones on their peripheral target tissues is not the prime concern of this book.

(i) *Oxytocin and Vasopressin*

Oxytocin and vasopressin elicit a variety of biological responses in addition to the classical activities described (see Section 3:7). Whilst the principal oxytocic actions are on the uterus and mammary gland stimulating smooth muscle contractions, this hormone also increases glucose oxidation by adipose (fat) cells. The major actions of vasopressin are to increase water permeability in the collecting ducts of the kidney and vasoconstriction of peripheral blood vessels although it can also increase glycogen breakdown by the liver. With respect to the pressor activity of vasopressin, this action is generally not considered to be important in the regulation of blood pressure, although its effects on blood vessels in the kidney and other splanchnic regions may be significant. In addition, vasopressin is active in stimulating the release of corticotrophin (ACTH) from the anterior pituitary and, with the anatomical evidence for vasopressin neurones terminating in the median eminence, there has been considerable speculation as to whether vasopressin may be a corticotrophin-releasing hormone (see Chapter 5 Section 3). Vasopressin has recently been identified in other neurones in the brain, outside the hypothalamus, where it has been implicated with learning and memory processes (see Chapter 11 Section 6). Among all the effects of neurohypophyseal hormones, only the actions of oxytocin on the uterus and mammary gland and the antidiuretic action of vasopressin can be elicited by administration of physiological doses of the hormones (i.e. levels equivalent to those measurable in the blood) whereas other biological effects of oxytocin and vasopressin (with the possible exception of effects in the central nervous system) can only be induced by supraphysiological doses of the hormones.

Peptide hormones are generally thought to bring about their specific activities by interaction with specialised parts of the membranes of their target cells, called 'receptors'. The only vasopressin receptors clearly identified are in the collecting ducts of the kidney of various species. Binding of radioactively labelled vasopressin occurs in membrane preparations of renal medulla and the receptors show a dissociation constant in the range of 10^{-9}–10^{-10}M (dependent on species) with a maximal binding capacity in the order of pmol per mg of protein. In other words, these results indicate a sensitivity to very low concentrations of hormone. It is well established that the antidiuretic effects of vasopressin in the kidney are mediated via cyclic AMP produced in the target cells (this intracellular second messenger is common to many cells regulated by hormones, see Chapter 1 Section 10). Vasopressin-sensitive adenylate cyclase (the enzyme which converts ATP into cyclic AMP) is located in the basal membrane (facing the blood) of tubular cells in the collecting ducts of the kidney tubule. The vasopressin-stimulated adenylate cyclase causes a rise in intracellular cAMP which ultimately leads to an increased water permeability at the luminal cell membrane (in contact with tubular fluid). This results in water reabsorption due to the osmotic gradient between fluid in the collecting ducts and the interstitial fluid surrounding the loops of Henlé. From experimental data it has been estimated that an occupancy of less than 5 per cent of vasopressin receptors results in maximal antidiuresis indicative of a large receptor reserve. High-affinity binding sites for [^{3}H]oxytocin with a dissociation constant of

1.8×10^{-9}M have been demonstrated in isolated mammary gland cells. The fact that the first biochemical events triggered by such binding are unknown has rendered complete characterisation of the receptors difficult, but again a high sensitivity of target cells to oxytocin is indicated.

(ii) *Roles of Secreted Neurophysins*

As described in Section 3:4, there is good evidence consistent with the concept of one neurone, one hormone, one neurophysin in the neurohypophyseal system, and generally factors which regulate hormone secretion also lead to release of neurophysins. It is perhaps surprising therefore that some investigators have found a relative excess of neurophysin in plasma when compared to vasopressin but these differences may be due to differential rates of degradation of the compounds – the half life of vasopressin is under 6 min whereas that for neurophysin is about 18 min. Nevertheless the recent development of radioimmunoassays which are specific for either vasopressin–neurophysin or oxytocin–neurophysin have shown that a stimulus which preferentially releases one or other of the hormones results in a measurable increase in the associated neurophysin with no increase in the alternative neurophysin.

Apart from their role as carrier proteins for the posterior pituitary hormones, there is some evidence that neurophysins may have an independent action since plasma levels of neurophysin may be elevated while the levels of vasopressin and oxytocin are not known to increase. This situation occurs in renal failure and during oestrogen administration. Some studies have implicated that certain neurophysins may act as a natriuretic factor (i.e. stimulate reabsorption of sodium in the kidney) whilst others may act as a lipolytic factor (i.e. release fatty acid from adipose tissue); but while it seems possible that some may possess biological activity no clear physiological function has been established for the neurophysins after secretion from the posterior lobe. However, the study of neurophysins has been enhanced by the development of specific and sensitive radioimmunoassays and further experiments on the secretory patterns of these carrier proteins may give new insights regarding the roles of secreted neurophysins. Recently it has been shown that vasopressin–neurophysin is also secreted into the hypophyseal portal system in both rat and monkey. The possible role for vasopressin as a corticotrophin releasing hormone has been alluded to, but a hypophysiotrophic role for the neurophysin, if any, has yet to be identified. For further reading on the roles of vasopressin in antidiuresis and cardiovascular homeostasis see [5, 9] and [12] in Chapter 11.

References

1. Acher, R. (1979) Neurophysins: molecular and cellular aspects. *Angewandte Chemie*, 18: 846-60.
2. Cross, B. A., Dyball, R. E. J., Dyer, R. G., Jones, C. W., Lincoln, D. W., Morris, J. R. & Pickering, B. T. (1975) Endocrine Neurones. *Recent Progress in Hormone Research*, **31**: 243-94.
3. Defendini, R. & Zimmerman, E. A. (1978) The Magnocellular Neurosecretory System of the Mammalian Hypothalamus, in S. Reichlin, R. J. Baldessarini & J. B. Martin (eds) *The Hypothalamus* (Raven Press, New York), pp. 137-52.
4. Dreifuss, J. J., Mathison, R. D. & Gamwilson, B. H. (1980) Secretion of Neurohypophyseal Peptides, in J. L. Barker & T. G. Smith (eds) *The Role of Peptides in Neuronal Function* (Marcel Dekker Inc., New York).
5. Gardiner, S. M. & Bennett, T. (1981) *Cardiovascular Homeostasis: Intrarenal and Extrarenal Mechanisms*. Oxford University Press.
6. Hayward, J. N. (1975) Neural control of the posterior pituitary. *Annual Review of Physiology*, **37**: 191-210.
7. Maddrell, S. H. P. & Nordmann, J. J. (1979) *Neurosecretion. Tertiary Level Biology*. Blackie.
8. McKelvy, J. R. & Epelbaum, J. (1978) Biosynthesis, Packaging, Transport and Release of Brain Peptides, in S. Reichlin, R. J. Baldessarini & J. B. Martin (eds) *The Hypothalamus* (Raven Press, New York), pp. 195-212.
9. Moses, A. M. (1980) Diabetes Insipidus and ADH Regulation, in D. T. Krieger & J. C. Hughes (eds) *Neuroendocrinology – a Hospital Practice Book* (Sinauer Assoc. Inc., Sunderland, MA), pp. 141-8.
10. Pickering, B. T. (1978) The neurosecretory neurone: a model system for the study of secretion. *Essays in Biochemistry*, **14**: 45-81.
11. Robinson, A. G. (1980) Neurophysis and their physiologic significance, in D. T. Krieger & J. C. Hughes (eds) *Neuroendocrinology – a Hospital Practice Book* (Sinauer Assoc. Inc., Sunderland, MA), pp. 149-58.
12. Russel, J. T., Gainer, H. & Brownstein, M. J. (1980) Biosynthesis of the common precursors of vasopressin, oxytocin and their respective neurophysins, in J. L. Barker & T. G. Smith (eds) *The Role of Peptides in Neuronal Function* (Marcel Dekker Inc., New York).
13. Schwartz, J. H. (1980) The Transport of Substances in Nerve Cells. *Scientific American*, **242** (4): 122-35.
14. Thorn, N. A., Russel, J. T., Torp-Pedersen, C. & Treiman, M. (1978) Calcium and Neurosecretion. *Annals of the New York Academy of Sciences*, **307**: 618-39.
15. Vincent, J.-D., Poulain, D. & Arnauld, E. (1980) Electrophysiology of Magnocellular Neuroendocrine Cells, in J. L. Barker & T. G. Smith (eds) *The Role of Peptides in Neuronal Function* (Marcel Dekker Inc., New York), pp. 229-72.

4

INTERMEDIATE LOBE OF THE PITUITARY: MSH SECRETION

Until recently, the pituitary intermediate lobe or pars intermedia has been more thoroughly studied in lower vertebrates than in mammals. This is because of its established role in the regulation of skin colour in fish, amphibia and reptiles, an effect which is mediated by the secretion of a hormone called melanocyte-stimulating hormone (MSH) which rapidly activates dermal chromatophores (see [1, 6] for example). Revived interest in the mammalian intermediate lobe has occurred as a result of the discovery that MSH and other peptides are derived from a pro-opiocortin (pro-opiomelanocortin) precursor in the intermediate lobe similar to that for the ACTH-related peptides of the anterior pituitary (see Figure 5.4) and the hypothalamus (see Chapter 10 Section 2). This chapter will outline our present understanding of the production, secretion and possible roles for α-MSH in the mammal together with a brief account of the significance of other forms of MSH and other peptides in the intermediate lobe, such as corticotrophin-like intermediate lobe peptide (CLIP), β-endorphin and β-lipotrophin (β-LPH).

4:1 Structure and Cell Types of the Intermediate Lobe

In many mammals, such as the cow, dog, rabbit and rat, the intermediate lobe is a distinct structure composed of several layers of epithelial cells, but, in other species it is often physically part of the neural lobe; hence, the term the neuro-intermediate lobe. In some mammals (e.g. whale, elephant and human) it is difficult to discern at all in the adult, but in the human foetus there is a large distinct intermediate lobe. The lobe consists of granular secretory cells which stain with anti-MSH antibodies but sometimes smaller corticotrophic-like cells can also be detected and these are usually associated with the pars distalis border (see Figure 1.3). The granular cells are surrounded by interstitial or stellate cells and, unlike the pars distalis, are richly innervated by both neurosecretory and catecholamine-containing nerve fibres which arise from cell bodies in the hypothalamus and reach the gland via the pars nervosa. Generally the neurosecretory neurones terminate near the pars nervosa whilst the aminergic neurones occur throughout the pars intermedia; this arrangement of neurosecretory and catecholamine nerve terminals is pertinent to the main mechanisms of regulation of MSH secretion (see Section 3). The gland is characterised by the presence of extracellular follicles containing colloid and, in this way, somewhat resembles the thyroid gland. However, the colloid does not appear to contain material secreted by the gland and its function remains to be determined. The blood supply to the intermediate lobe is mostly via the pars nervosa vessels and partly from the hypophyseal portal system, but the poorly defined intermediate lobe is also poorly vascularised in many mammals, including rodents, primates and man. For further details on the structure of the pars intermedia see [5, 6] and H. Heller & K. Lederis (eds) (1971) *Subcellular Organisation and Function in Endocrine Tissues* (Cambridge University Press).

4:2 Melanocyte Stimulating Hormones (α-MSH, β-MSH and γ-MSH)

(i) *Structures and Derivation*

Three types of MSH have been isolated from pituitary gland extracts – α-MSH (13 amino acids), β-MSH (18 amino acids) and γ-MSH (12 amino acids). The structures of α- and β-MSH have been known since 1960 although only the secretion of α-MSH has been proven and well studied in a number of species but not in the adult human. β-MSH may not be a true hormone but simply a breakdown of β-lipotrophin (β-LPH) secretion (see Figure 5.4) γ-MSH was described very recently as an *N*-terminal sequence present in pro-opiocortin and its possible role as a hormone is also unknown. α- and β-MSH share a sequence of seven amino acids (4 to 10) with ACTH and γ- and β-lipotrophin which explains the MSH-like activity of these peptides. MSH biological activity can be measured by determining changes in skin colour when administered to amphibia or other lower vertebrates but sensitive radioimmunoassays for α-MSH and other related peptides are now in use and have provided a better analysis of the different forms of MSH. In the pars intermedia, pro-opiocortin is processed to an ACTH-like peptide and approximately 20 per cent of rat pituitary ACTH occurs in the intermediate lobe (Figure 5.2). It is possible that some of this ACTH is secreted from the intermediate lobe in some species but it is generally held that the granular cells of the intermediate lobe process ACTH by enzymic tryptic cleavage at the 16–17 peptide bond. The *N*-terminal fragment forms α-MSH by the action of a carboxypeptidase to remove three amino acids followed by *C*-terminal amidation and *N*-terminal acetylation (Figure 4.1) and the *C*-terminal fragment of

Figure 4.1: Pars Intermedia Hormones: Formation of α-Melanocyte Stimulating Hormone (see also Figure 5.2 and 5.3)

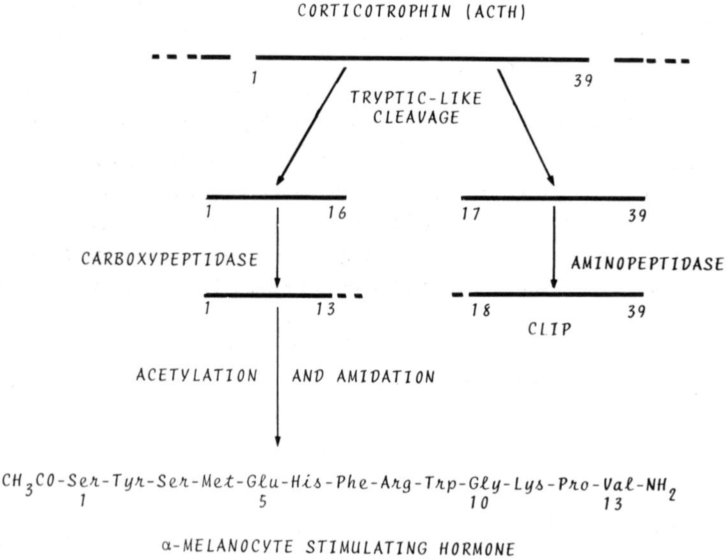

ACTH forms CLIP (ACTH 18–39) by the action of an aminopeptidase. Similarly the granular cells of the intermediate lobe process the *C*-terminal pro-opiocortin or β-LPH to γ-LPH and β-endorphin. The pro-opiocortin precursor is a glycoprotein (a protein with attached sugar molecules) and it has been shown that the presence of the sugar is critical for normal processing of peptides from the precursor and may be used for regulating such mechanisms.

(ii) *Actions of MSH*

The main pro-opiocortin peptides characterised from the corticotrophic cells of the anterior pituitary in all species studied are ACTH, β-LPH and pro γ-MSH and no further processing of these peptides appears to occur here. This contrasts with the intermediate lobe where there are interspecies differences in the amounts and numbers of peptides processed. Generally, higher mammals, particularly the primates including man, have a less active intermediate lobe than that of lower forms such as rodents. This difference has led to some discussion regarding the possible roles of MSH and other peptides in mammals and indeed, whether or not intermediate lobe peptides, with the exception of α-MSH, are actually secreted or have any peripheral function in humans and other mammals. These peptides, however, are produced in the brain in higher mammals suggesting that they may have some central action[3, 4]. Evidence for a number of actions for α-MSH in the rat have indicated possible roles in some mammals. α-MSH, acting synergistically with androgens, is known to stimulate sebum secretion from the sebaceous glands in the skin and also the secretion of pheromones from specialised sebaceous glands (the preputial glands) which are associated with sexual behaviour in this species (see Chapter 11 Section 8 and [7]). It is also possible that α-MSH may have particular adrenal functions in the regulation of cortisol and aldosterone secretion particularly in the foetal and neonatal animal. In fact the presence of a large pars intermedia in foetal primates suggests that pars intermedia peptides may have important functions before or around birth in higher mammals, but have no significant function in the older animal when the gland atrophies.

4:3 Neural Factors Determining α-MSH Secretion

The secretion of α-MSH from the intermediate lobe is under central inhibitory control by the hypothalamus [5] very much analogous to the control of prolactin secretion. If the gland is disconnected from the brain or the hypothalamic control is inhibited in some way, α-MSH secretion is increased and hormone content is temporarily reduced. In the rat, dopaminergic neurones originate in the arcuate nucleus of the hypothalamus (see Figure 1.2) and studies with dopamine receptor agonist and antagonist drugs indicate that pars intermedia cells are directly inhibited by dopamine. Such studies suggest that hypothalamic dopamine neurones innervate the cells of the intermediate lobe and inhibit the secretion of α-MSH (Figure 4.2).

Further evidence for an inhibitory dopaminergic control of intermediate lobe secretions exerted by the hypothalamus comes from *in vitro* studies on isolated intermediate lobe cells held in a biogel column (see Chapter 2 Section 3(v)). Dopamine inhibits the secretion of all intermediate lobe peptides from these cells. These were measured by radioimmunoassay and include α-MSH, ACTH, γ-LPH and β- and γ-endorphin. This inhibition of all peptides is not surprising since they are derived from a common precursor and are probably stored in the same secretory granules, but the physiological significance of this control is uncertain. Exogenous adrenaline, noradrenaline, γ-aminobutyric acid and 5-hydroxytryptamine also modify the release of α-MSH and ACTH from neurointermediate lobe cells, and these transmitters may be involved in the neural control of the neurointermediate lobe (Figure 4.2). Recent studies have shown that isoprenaline (a β-adrenoreceptor agonist) is particularly potent on α-MSH and ACTH release from the rat neurointermediate lobe further implicating a noradrenergic control of secretion.

Apart from the aminergic innervation of the pars intermedia, neurosecretory peptidergic neurones from the paraventricular nucleus in the hypothalamus directly innervate the intermediate lobe, but marked species differences exist concerning the degree of this innervation. In rodents, for example, the neurosecretory terminals which penetrate the intermediate lobe are fewer than in other mammals. It is assumed the peptidergic control involves two hypothalamic factors, one which stimulates (MSH-releasing factors, MRF) and the other which inhibits (MSH inhibiting factor, MIF) the release of α-MSH from the pars intermedia and these activities are present in extracts of hypothalamic tissue[3]. Earlier reports, which have not been confirmed recently, suggested that these factors may be derived from oxytocin, since the *C*-terminal tripeptide of oxytocin (see Figure 3.1), Pro-Leu-Gly-NH_2 shows MIF activity, whereas H-Cys-Tyr-Ile-Gly-Asn-OH acts as an MRF. Such a possibility is enhanced by cytochemical evidence that

Figure 4.2: Control of α-MSH Secretion: Possible Factors Involved. MSH – melanocyte stimulating hormone (melanotrophin); MIF – MSH inhibitory factor; MRF – MSH releasing factor; OXY – oxytocin; DA – dopamine; A – adrenaline; NA -- noradrenaline; GABA – γ-amino butyric acid; 5-HT – 5-hydroxytryptamine; ARC – arcuate nucleus; PVN – paraventricular nucleus.

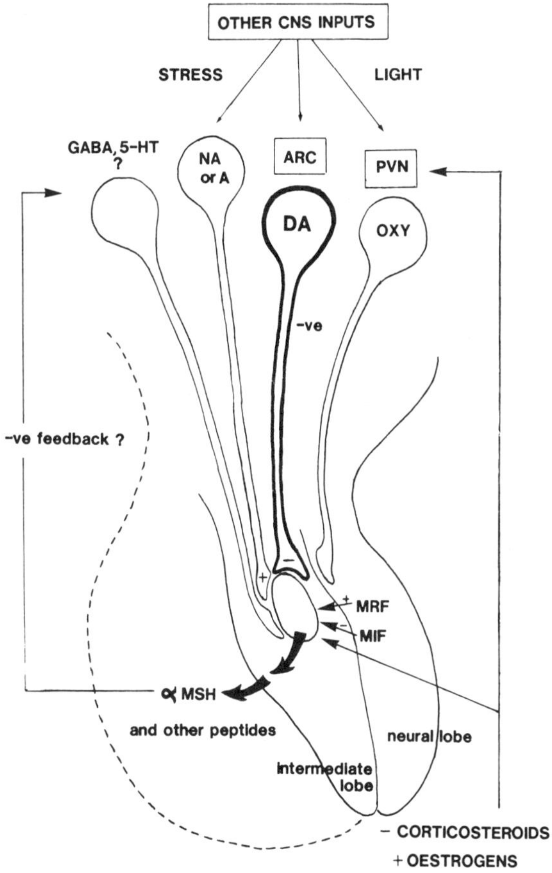

oxytocin-containing nerve fibres and terminals in the neural lobe occur close to the pars intermedia border (Figure 4.2). In conclusion, the hypothalamic inhibitory control of MSH secretion involves a complex but direct neural mechanism which contrasts the neurohumoral control system of the anterior pituitary. Further understanding of the nature of the integration of the catecholaminergic and neurosecretory components of this control depends upon future research into these mechanisms.

A number of environmental factors influence MSH secretion in lower vertebrates and appear to have similar effects in mammals although their physiological significance in mammalian species remains uncertain. Stressful stimuli generally stimulate MSH secretion and increase MSH synthesis in the intermediate lobe. Light inhibits MSH secretion in amphibia and fish and, although the evidence is sparse and confusing, light may also affect the release of MSH from the mammalian intermediate lobe. Some evidence also suggests that MSH secretion is also altered by auditory and tactile inputs in the rat and by suckling stimuli in lactating animals. A further factor which appears important in mammalian neurointermediate lobe function, for reasons which are not clear, is the availability of water and the animals' state of hydration. Administration of hypertonic saline induces MSH secretion in rats, and other rodent species which dwell in hot, desert regions usually have a well-developed intermediate lobe.

For further information on the control of MSH secretion see [3, 5].

4:4 Endocrine Factors and Feedback Control of α-MSH Secretion

Experiments have implicated that MSH may regulate its own secretion by a short-loop feedback mechanism (see Chapter 1 Section 9) in both lower vertebrates and mammals. Also sporadic and somewhat conflicting evidence exists that a number of peripheral hormones may feedback and modulate MSH secretion – foremost amongst these are the oestrogens and progesterone. Measurement of pituitary α-MSH content and release throughout the rat oestrous cycle and studies of steroid treated-ovariectomised rats generally indicate changes consistent with stimulation of synthesis and secretion of α-MSH by the gonadal steroids. Also in the rat changes in corticosteroids that alter the secretion of ACTH may also affect α-MSH and ACTH release from the intermediate lobe. However, alternative evidence indicates that adrenalectomy or cortisol treatment has little or no effect on pituitary content or blood levels of α-MSH. Furthermore, changes in either thyroid or pineal function have been reported to modulate α-MSH secretion but more experiments are needed to confirm this.

4:5 Other Intermediate Lobe Peptides

(i) *Corticotrophin-like Intermediate Lobe Peptide (CLIP)*

CLIP ($ACTH_{18\text{-}39}$) is formed in the intermediate lobe and may be secreted along with α-MSH in rat and other species, although neither peptide has been detected in adult human blood. No clear physiological role for CLIP has so far been defined but in rodents this peptide and/or smaller *C*-terminal fragments may be involved with the stimulation of insulin secretion. In humans recent work has suggested that foetal pituitary contains large quantities of α-MSH and CLIP-like activity but from the time of parturition mostly ACTH is present in the pituitary and little α-MSH or CLIP. Based on this and the evidence that α-MSH stimulates the foetal adrenal gland, it is possible that α-MSH and CLIP may be important regulators of adrenal function in the foetus; future research will no doubt substantiate or refute these possible roles for CLIP and α-MSH.

(ii) *γ-Lipotrophin and β-Endorphin*

These two peptides are processed from pro-opiocortin via β-LPH in the intermediate lobe of various mammalian species (see Figure 5.2). Little is known about γ-LPH but it has been detected in adult human pituitary, blood and cerebrospinal fluid. β-endorphin is formed in both the intermediate lobe and anterior pituitary but the secretion of intact and biologically active β-endorphin from either region remains uncertain. More β-endorphin occurs in the intermediate lobe but some studies suggest that in this region the peptide is effectively inactivated by proteolysis to α-endorphin or acetylated possibly by the enzyme system which forms α-MSH. In the anterior pituitary corticotrophs, ACTH and β-lipotrophin are the main peptides formed but β-endorphin is also present and possibly secreted intact. β-endorphin induces profound analgesia in the central nervous system (see Chapter 10 Section 3) and pituitary β-endorphin may have important analgesic effects following its co-release with ACTH during stress and in response to corticotrophin releasing factor. Although the blood concentrations of β-endorphin which occur in stress may not be high enough to induce analgesia, β-endorphin may also have important endocrine effects as it has been shown to modulate the secretion of a number of pituitary hormones (see Table 10.4) and affects intermediary metabolism particularly through its actions on the pancreas and on eating behaviour[2].

For further reading there is an excellent up-to-date review of 'the peptides of the pars intermedia', Ciba Foundation Symposium 81

(Pitman Medical Ltd).

References

1. Baker, B. I. (1979) The Evolution of ACTH, MSH and LPH-structure, Function and Development, in E. J. W. Barrington (ed.) *Hormones and Evolution* (Academic Press, New York), pp. 643-722.
2. Imura, H. & Nakai, Y. (1981) 'Endorphins' in Pituitary and Other Tissues. *Annual Review of Physiology*, **43**: 265-78.
3. Kastin, A. J., Schally, A. V. & Kostrzewa, R. M. (1980) Possible Aminergic Mediation of MSH Release and of the CNS effects of MSH and MIF-I. *Federation Proceedings*, **39**: 2931-6.
4. Krieger, D. T., Liotta, A. S., Brownstein, M. J. & Zimmerman, E. M. (1980) ACTH, β-LPH and Related Peptides in Brain, Pituitary and Blood, *Recent Progress in Hormone Research*, **36**: 277-337.
5. Taleisnik, S. (1978) Control of Melanocyte Stimulating Hormone (MSH) Secretion, in S. L. Jeffcoate & J. S. M. Hutchinson (eds) *The Endocrine Hypothalamus* (Academic Press, New York), pp. 421-39.
6. Turner, C. D. & Bagnara, J. T. (1971) Pars Intermedia: Chromatophore Regulation among Vertebrates, in *General Endocrinology* (W B Saunders), pp. 135-61.
7. Thody, A. J. (1977) The Significance of Melanocyte Stimulating Hormone (MSH) and the Control of its Secretion in the Mammal. *Advances in Drug Research*, **2**: 23-74.

5

CONTROL OF CORTICOTROPHIN SECRETION

5:1 The Hypothalamic-Pituitary-Adrenal Axis

The secretion of adrenocorticotrophin (ACTH) from the anterior pituitary is essential for maintaining the secretory activity of the adrenal cortex. ACTH stimulates the synthesis and secretion of the glucocorticoids, corticosterone and cortisol (hydrocortisone) (Figure 5.1), the latter being the principle steroid secreted by the majority of mammals although in rodents, such as the mouse and rat, corticosterone is the predominant hormone. These corticosteroids are secreted during stress and exert powerful 'anti-insulin' effects on both carbohydrate and protein metabolism. Glucocorticosteroids induce several enzymes in liver concerned with the synthesis of glucose from amino acids, and also enzymes in tissues, such as muscle, which breakdown protein and release amino acids. Corticosteroids and also ACTH/β-LPH-like peptides (see Figure 5.2) stimulate lipolysis in adipose (fat) tissues although the importance of the peptides in regulation of fat metabolism is uncertain. In addition, glucocorticosteroids in high, supraphysiological concentrations markedly suppress immune responses and are therefore beneficial when given to patients during and after surgical transplant or tissue graft operations; but large concentrations also inhibit wound healing and more general inflammatory reactions which occur as a response to allergic conditions such as hay fever and rheumatoid arthritis. Perhaps the most important physiological functions of cortisol, however, involve its homeostatic and permissive roles which

Figure 5.1 Synthesis of Steroid Hormones.

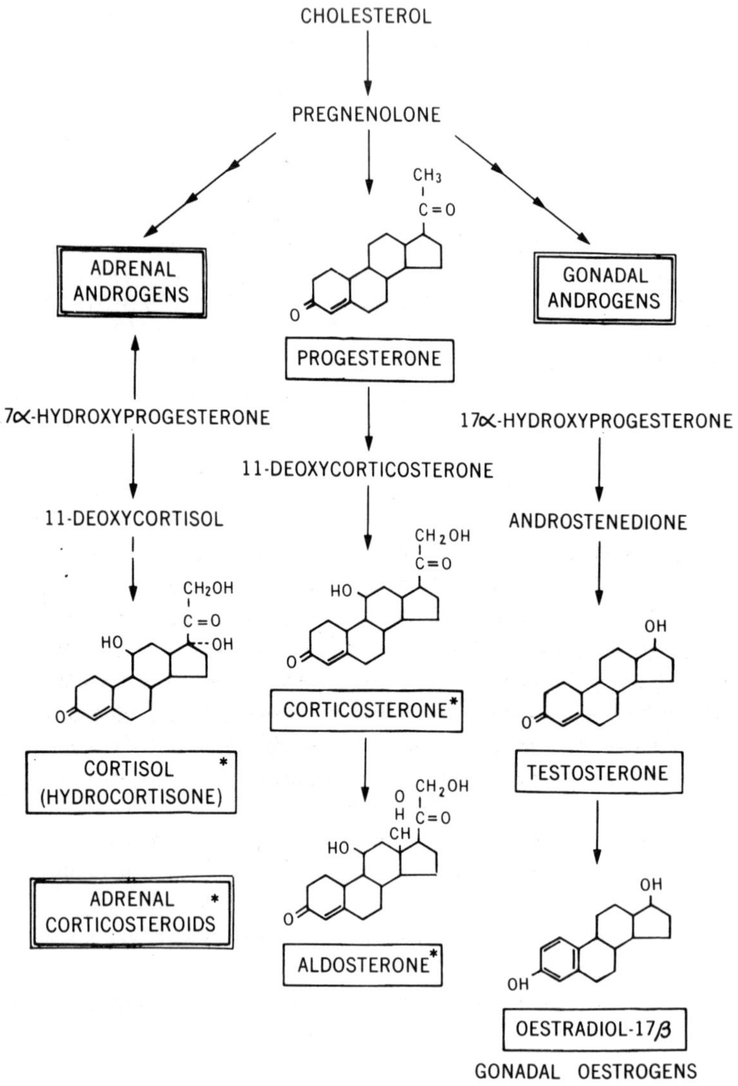

* Indicate adrenal corticosteroids

Figure 5.2: The Structural Relationships of ACTH and Related Peptides Indicating Sequence Homologies and Possible Derivations.

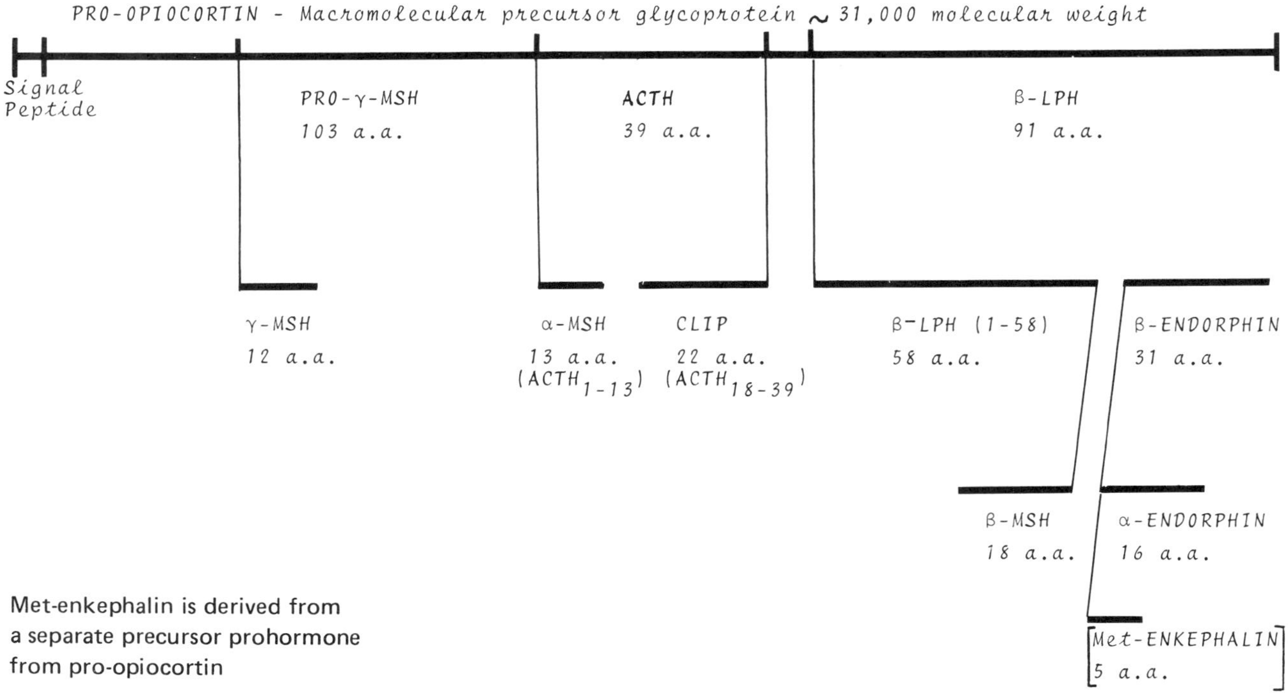

collectively protect the organism from stress. Finally, specific corticosteroids act on pituitary and brain (see Section 5:6) and alter behaviour (see Chapter 11 Section 2). Mineralocorticosteroids, particularly aldosterone, stimulate sodium reabsorption in the kidney. For further details of the diverse physiological effects of corticosteroids see V. H. J. James & J. R. Pasqualini (eds) (1980) *Hormonal Steroids* (Pergamon Press).

It has been known for more than 25 years that the secretion of ACTH is under rigorous control by specific factors or neurohormones released from the hypothalamus and that these substances are transported to the anterior pituitary by the hypophyseal portal vessels. The major central components of the hypothalamic pituitary adrenal axis are indicated in Figure 5.5 and it is the purpose of this chapter to review the current knowledge of the control mechanisms which regulate ACTH secretion. The effects of hypothalamopituitary dysfunction on the adrenal axis, together with information on the tests used to determine normal ACTH secretion in humans, are covered in Chapter 12 Section 2.

5:2 ACTH and Related Peptides

The cells which synthesise and secrete ACTH are called corticotrophs and are randomly distributed in the mammalian pars distalis. Histological studies utilising both light and electron microscopy have shown that the corticotrophs in the anterior pituitary are chromophobic and basophilic and contain relatively fewer and smaller secretory granules than neighbouring adenohypophyseal hormone-secreting cells (see Plate 2). It is also evident from studies in a number of species, including humans, that corticotrophs co-synthesise and secrete the protein β-lipotrophin (β-LPH) and possibly further related peptides (see Figure 5.2).

(i) *Structure of ACTH*

The discovery of a pituitary hormone which regulates steroid production by the adrenal cortex is accredited to the work of P. E. Smith in the 1920s although it was not until the 1950s that this hormone was finally isolated and structurally characterised by P. M. Bell and his collaborators. Applying similar biochemical techniques to those used for the chemical identification of the posterior pituitary hormones (see Chapter 3 Section 1), they described a linear polypeptide of 39

Figure 5.3: Structure of Human ACTH.

```
NH2-Ser-Tyr-Ser-Met-Glu-His-Phe-Arg-Trp-Gly-Lys-Pro-Val
     1               5                  10              Gly
                Tyr-Val-Lys-Val-Pro-Arg-Arg-Lys-Lys
             24 Pro
                Asn-Gly-Ala-Glu-Asp-Glu-Ser-Ala-Glu-Ala-Phe-Pro-Leu-Glu-Phe-COOH
                25                 30                  35                39
```

amino acid residues. However, several years elapsed before the 1-39 ACTH was synthesised and its structure thus validated (Figure 5.3). ACTH has a similar structure in all mammalian species studied with only small variations occurring in the 25-33 sequence of amino acids but it was shown in the earliest work that the biological activity of the hormone resides in the *N*-terminal 1-24 amino acids and that the *C*-terminal sequence (25-39) is not essential for full steroidogenic activity of the adrenal cortex.

(ii) *The ACTH Precursor Molecule Pro-opiocortin*

Our understanding of the synthesis, secretion and biological function of ACTH has developed rapidly during the 1970s. An important development which has precipitated a major re-evaluation of many earlier studies of ACTH regulation was the discovery that this hormone is produced from part of a macromolecular precursor protein or glycoprotein, pro-opiocortin (Figure 5.2). This precursor not only contains the ACTH sequence but also a larger polypeptide component, β-lipotrophin (β-LPH) that is co-secreted from identical granules and may function as a hormone in its own right. In addition, as the sequence homologies in Figure 5.2 indicate, both ACTH and β-LPH may serve as precursors for other biologically active peptides or hormones, i.e. β-endorphin, α-melanocyte stimulating hormone (α-MSH) and corticotrophin-like intermediate lobe peptide (CLIP). These homologous relationships have been determined and confirmed by the isolation and chemical characterisation of the peptide sequences and also by immunological studies using highly specific antibodies for different peptide sequences. Pituitary production of an ACTH precursor has been recently confirmed by identifying a complete DNA sequence which contains a duplex copy of an ACTH-β-LPH precursor in RNA molecules.

These findings have, naturally, created a great deal of discussion by scientists working in this area. Although the position is by no means resolved at the present time, the intriguing implications are that the large pro-opiocortin contains yet further peptide sequences which are processed and co-secreted with hormones, such as ACTH, and that these peptides may act as co-hormones regulating related functions in the adrenal cortex and other sites. Some preliminary evidence for this has already emerged. The opiocortin precursor has been isolated from the adenohypophysis of a number of species and is common to all tissues which produce any one or more of this family of peptides, namely the anterior pituitary, intermediate lobe, and various brain regions, including the hypothalamus (see Chapter 10 Section 2) as well as parts of the alimentary tract and the placenta.

What is reasonably well established is that ACTH and β-LPH are elaborated and released from the anterior pituitary corticotrophs, whereas α-MSH and CLIP are processed from ACTH precursors in the intermediate lobe although ACTH itself may be secreted from certain pars intermedia cells in rat and dog but not all species (see Chapter 4 Section 2). Earlier studies suggested that in the human, ACTH and β-melanocyte stimulating hormone (β-MSH) were secreted concomitantly, but β-MSH was later shown to be a breakdown product of β-LPH, thus indicating possible problems of interpretation using immunological assays (see Chapter 5 Section 2(iii)). However, recent evidence has confirmed that β-endorphin is formed in and secreted from the anterior pituitary although its function remains uncertain (see Chapter 4 Section 5). In addition, recent sequence studies of pro-opiocortin indicate a further *N*-terminal MSH-like peptide (pro-γ-MSH) (see Chapter 4 Section 2).

The various peptide fragments which are formed are processed from their precursors by the action of specific proteolytic enzymes (peptidases) in the different tissues, some of which have been identified. It is particularly interesting to observe the amino acids at the proposed points of peptide cleavage in the precursor molecule; these sites are invariably flanked by paired basic amino acids (e.g. Arg-Arg- or Lys-Lys) and such paired residues are points at which proteolytic enzymes are known to act. For further information on pro-opiocortin see [5].

(iii) *Measurement of ACTH*

Up until the mid 1960s measurement of ACTH in tissues, blood and other body fluids was carried out by bioassays which often had limited sensitivity and specificity. For *in vitro* bioassays the amount of corticosteroid released by isolated adrenal tissue in response to ACTH was measured either fluorimetrically or by the ACTH-induced reduction in adrenal ascorbic acid. Alternative *in vivo* assays involved the assessment of adrenal corticosteroid secretion in response to administered ACTH in animals either hypophysectomised or drug or steroid pre-treated to remove endogenous ACTH secretion.

The development of radioimmunoassay procedures by Benson, Yalow, Greenwood and others in the 1960s resulted in a wide application of this method to hormone measurements including ACTH. However, since all immunological procedures depend on the unique affinities of the antisera used, such methods have been criticised since they do not distinguish immunoreactive fragments of the hormone which may not be biologically active. In other words the assays are prone to 'false-positive' measurement. More recent techniques with high sensitivity and biological specificity are the cytochemical assays (see Chapter 2 Section 3(iv)). Such assays were amongst the first to enable biological measurement of the lowest resting levels of ACTH in small samples of human or rat plasma. These methods, however, are perhaps less appropriate for routine bulk measurement of samples as required in clinical or research laboratories, and for this reason immunological assays for ACTH are widely used. For further information on ACTH see [9].

5:3 Corticotrophin Releasing Factor (CRF)

In the early 1950s increasing attention was being paid to the pioneer studies of G. W. Harris and others which suggested a neurohumoral regulation of anterior pituitary function. It was M. Saffran and colleagues however, who first demonstrated that the hypothalamus and neurohypophysis contain substances which stimulate ACTH release and they coined the name corticotrophin releasing factor (CRF). Subsequently similar studies confirmed this activity in the absence of measurable amount of other agents known to modulate ACTH secretion. At around the same time, Porter and colleagues showed that hypophyseal portal blood from hypophysectomised, stressed dogs contained a potent CRF activity.

(i) *Structure of CRF*

It is well over a decade since the first hypothalamic releasing factor, thyrotrophin releasing hormone (TRH) was isolated and chemically characterised by the independent work of A. V. Shally and R. Guillemin (see Chapter 6 Section 3). Since that time several hypothalamic releasing factors and other neuropeptides have been characterised but the nature and structure of CRF still remains uncertain, despite intensive investigations spanning nearly a quarter of a century. The reasons for this are associated, firstly, with the problem of extracting and assaying a relatively labile substance which is non-specifically released during most forms of stress, and secondly, the finding that several substances other than CRF, (such as vasopressin) can induce ACTH release.

In the early 1960s A. V. Schally and his colleagues partially identified two CRF-like substances one of which resembled vasopressin and the other α-MSH. However, numerous experiments suggested that these peptides could not solely account for the CRF activity in median eminence extracts. In an attempt to circumvent lengthy tissue extraction procedures, which could chemically transform the CRF during extraction, the CRF released from incubated hypothalamic tissue has been collected and partially characterised. The CRF was released by 5-hydroxytryptamine (5-HT) which does not evoke vasopressin release (see Sections 5:3(v) and 5:7). Again two separate peptidergic peaks of CRF activity were found but the amino acid sequence of these partially purified peptides has yet to be determined. Despite earlier difficulties a linear polypeptide of 41 amino acids has been isolated very recently from side fractions of the sheep hypothalamic extracts previously used to isolate the gonadotrophin releasing hormone (Gn-RH). This large peptide releases both ACTH and β-endorphin from rat pituitaries. Although it is too early to predict with any certainty, it is now possible that the nature of the elusive CRF molecule has finally been determined, and if so then this hypophysiotrophic factor may achieve the true status of an established hypophysiotrophic hormone[3]. On the other hand, future research may reveal yet further factors with CRF activity.

(ii) *Measurement of CRF*

Numerous bioassay procedures have been developed in an attempt to quantify CRF activity, all of which involve measurement of either ACTH secretion or adrenal corticosteroid production as an end point. Both *in vivo* and *in vitro* bioassays for the measurement of CRF have been developed. The former involves assessing ACTH secretion following administration of a CRF extract and this necessitates removing the influence of the endogenous CRF on the anterior pituitary either by lesions of the median eminence (where most CRF is stored), transplanting the pituitary away from the median eminence or blocking endogenous CRF release by drugs or corticosteroid treatment. Bioassays on isolated pituitary tissue are generally easier and more sensitive, and again involve measuring ACTH release in response to CRF extracts. Such *in vitro* assays are carried out on paired hemi-pituitaries (one half being used as control) or on isolated dispersed pituitary cells. Perhaps the most sensitive and specific *in vitro* assay that has been developed recently has utilised segments of rat pituitary which were 'primed' with vasopressin and ACTH secretion was subsequently measured by the highly sensitive cytochemical assay. For further information on CRF measurement see [2].

(iii) *CRF Neurones*

The discrete localisation of TRH, the gonadotrophin releasing hormone (Gn-RH) and somatostatin within the CNS has been defined by radioimmunoassay and immunohistochemical techniques. However, since the chemical structure of CRF is uncertain, we do not have CRF antibodies and thus the localisation of CRF can only be determined by measuring the concentration of CRF material by relatively unsatisfactory bioassay methods. Such studies have shown that the highest levels of CRF in the brain are found in the median eminence region although it is also present in other regions of the hypothalamus indicating a rather diffuse distribution of CRF-containing cell bodies. Microinjection of a CRF extract into rat hypothalamus has shown that when the material is injected into the posterior medial basal hypothalamus it is more effective in eliciting the release of ACTH than more anterior sites of injection, thus implicating a posterior localisation of CRF neurones.

As yet there is no evidence that CRF exists in other areas of the brain outside of the hypothalamus, as has been shown for other hypothalamic hormones such as TRH and somatostatin. However, this negative finding may simply reflect the insensitivity of the bioassay methods available to quantify the concentration of the peptide. It has been reported that CRF activity can be detected in the blood of animals long after removal of the hypothalamus. The confirmation of extrahypothalamic sources of CRF necessitates further experiments although it has been suggested that alternative sources may play a part in a 'back up' system in conditions of severe chronic stress (e.g. in injury or during surgery).

(iv) *Action of CRF on the Pituitary Corticotrophs*

It is generally held that the action of CRF on the pituitary corticotrophs, as with other hypophysiotropic hormones, is mediated by the intracellular messenger cyclic adenosine 3′:5′-phosphate (cyclic AMP) (see Chapter 1 Section 10). Cyclic AMP, its various derivatives and theophylline, (a drug that inhibits the enzyme phosphodiesterase which rapidly metabolises cAMP to AMP) all stimulate ACTH release. Furthermore, prostaglandins (see Chapter 1 Section 10) are thought to activate this process in some way. For further reading on CRF see [1, 2, 3].

(v) *Vasopressin as a CRF*

Vasopressin stimulates ACTH secretion in the intact animal and in isolated pituitary tissues; there is also a vasopressin neurophysin projection to the external zone of the median eminence (see Chapter 3 sections 2, 10) and thus vasopressin can be transported in the hypophyseal portal blood to the anterior pituitary corticotrophs. Despite this there is sufficient evidence that this hormone is not the main CRF since vasopressin and CRF activity are not always secreted at the same time, and partially purified hypothalamic CRF does not possess vasopressin activity. Although CRF and vasopressin are distinct entities, some relationship between them might be expected to exist. Several studies have suggested that vasopressin potentiates ACTH release by CRF at the level of the hypothalamic median eminence or the pituitary corticotroph. It has been suggested that the effect of vasopressin may be to help to override the negative feedback of corticosteroid hormones during stress. Interestingly however, Brattleboro rats, which are genetically deficient in hypothalamic vasopressin are still able to respond to stress by secreting ACTH, suggesting that the CRF role of vasopressin is not critical in this species. Angiotensin II stimulates vasopressin secretion (see Chapter 3 Section 9(i)) which may explain the reported CRF activity of this neuropeptide (see Figure 5.5, and Chapter 11 Section 9).

5:4 Neural Control of ACTH Secretion

(i) *Basal Secretion: Circadian Rhythm*

The chief factors regulating ACTH secretion operate through the hypothalamus and the release of CRF (see Figure 5.5). The basal secretion of ACTH has been shown to be episodic with periodic bursts of secretion occurring every one to three hours depending on the time of day. Measurement of blood plasma levels over a 24 hour period indicates a circadian rhythm of ACTH release in both human and animals. This occurs due to changes in CRF secretion resulting in an alteration in the periodicity of ACTH and thus daily fluctuations in

Figure 5.4: Circadian Variation in ACTH and Cortisol Levels in Human Plasma.

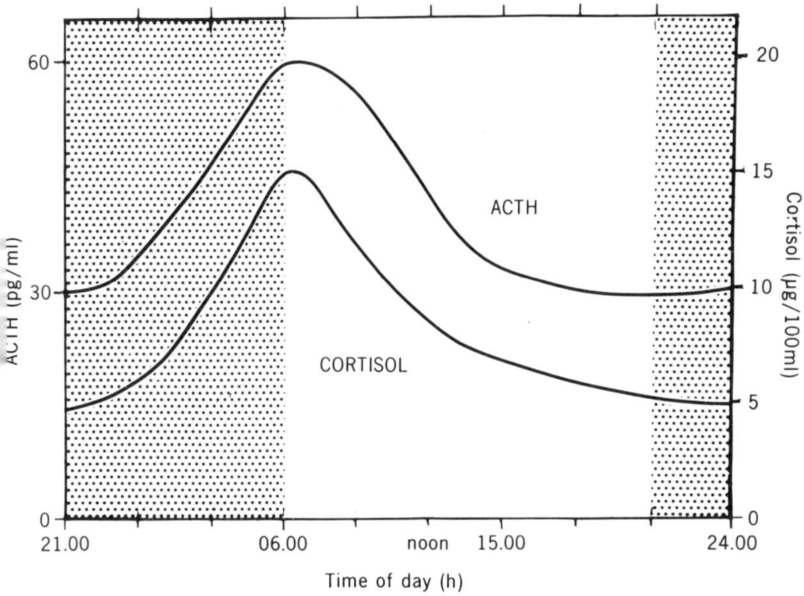

the mean levels of circulating ACTH and corticosteroids (see Figure 5.4). Such a circadian pituitary–adrenal rhythm is entrained to both dark/light and sleep/wakefulness cycles such that in humans, levels of circulating ACTH and cortisol peak in the morning before waking and trough in the evening before sleep. This rhythm is disrupted by changing the sleep pattern and in nocturnal animals (e.g. rat) the entire cycle is reversed so that peak levels of the hormones occur in the early evening.

In the rat, circadian changes in CRF correlate closely with changes in hypothalamic 5-hydroxytryptamine (5-HT). The 5-HT neurones emanate from the reticular activating system (RAS) in the brain stem which regulates sleep/consciousness states and the neurones partly terminate in the hypothalamic supra-chiasmatic nucleus which is highly implicated in the control of circadian rhythms (see Chapter 9 Section 2). Imposed on the diurual variation are the more marked changes in the hypothalamic-pituitary-adrenal axis induced by stress and the magnitude of this response to stress is subject to a circadian variation in that it is usually higher when resting plasma steroid levels are at their lowest.

(ii) *ACTH secretion in Response to Stress*

As early as the mid 1930s it was demonstrated by H. Selye and others that 'diverse nocuous agents' (i.e. various stressful stimuli) in rats produced a syndrome involving hypertrophy of the adrenal cortex and increased corticosteroid production: subsequently, the mediation of pituitary ACTH and hypothalamic CRF in this response was shown. As a result of complex neuronal inputs to the CRF neurones following stress, there is rapid synthesis and release of CRF from the rat median-eminence within 1–2 min which causes an immediate increase in pituitary ACTH synthesis and release. Within 3 to 4 min, blood levels of ACTH are elevated four- to five-fold. This ACTH stimulates the adrenal cortex and results within 10 min in an eight- to ten-fold increase in plasma levels of corticosterone in the rat. Similarly, emotional or physical stress in humans rapidly results in a three- to five-fold increase in plasma cortisol levels. These rapid changes in response to stress indicate the highly tuned nature of the hypothalamic pituitary axis and the large amplification of responses occurring from brain to adrenal glands. The increase in blood corticosteroid levels, well above the normally low resting levels, appears to be a vitally important component of the adaptive response to stress and to the survival of the organism. Animals which are either hypophysectomised or adrenalectomised (i.e. with their pituitary or adrenals removed) will not survive any chronic stressful treatment unless they are administered appropriate levels of corticosteroid. This long-term survival results from the multitude of actions which corticosteroids have on many body tissues. More immediate survival during stress is assisted by the increased output of adrenaline from the adrenal medulla, brought about by activation of the sympathetic component of the autonomic nervous system, and this acute physiological response to stress is perhaps more widely known and its mechanism better understood. For further information on adrenal medulla function and stress see [8].

Broadly speaking there are two categories of stressful stimuli although these are by no means mutually exclusive. Firstly 'neural stress' which involves a neural input to the CRF neurones and results from psychological or emotional responses and secondly systemic stress which results from inputs from somatic or visceral responses, namely metabolic or tissue changes including injury and responses to drugs. Such systemic stress may involve a neural input through the CRF neurones or may operate directly on the pituitary adrenal axis but such stress almost invariably incorporates a psychological component as well. One factor which complicates the study of neuroendocrine responses to stress is the extremely large variation observed between individuals of the same species. This is particularly true in the case of emotional stress and it should be noted that the degree of stress is most difficult to quantify. The neural pathways involved in 'emotional' and 'systemic' stress are clearly separable but their complex differentiation and integration in the hypothalamus is not well understood. Many physiological studies involving specific stimulation or lesion of brain regions have implicated both the amygdala and hippocampus in the regulation of CRF secretion. Together with the hypothalamus, these regions are important parts of the limbic system concerned with the control of emotional behaviour and thus it is not surprising to find that neural inputs from the hippocampus and amygdala to the hypothalamus may modulate CRF secretion. Similarly, neurophysiological experiments have demonstrated specific afferent pathways associated with certain forms of systemic stress (e.g. leg fracture or noxious sensory stimulation) and these multisynaptic pathways ascend in the spinal cord and, in common with the 5-HT neurones mentioned previously, pass through the brain stem, mid-brain and the thalamus before impinging on neurones in the hypothalamus. A number of brain neurotransmitters have been shown to alter CRF secretion (see Section 7) although in many cases it has not been possible to relate these to specific pathways which ultimately terminate on CRF neurones. For further reading on the neural control of ACTH see [4].

5:5 Feedback Control of CRF and ACTH Secretion by Corticosteroids

The feedback inhibition of adrenal corticosteroids on the release of ACTH is well established, but the sites and mechanism of action of this process have been, and still are, widely debated. The use of radioactive labelled corticosteroids have shown that central target cells which retain [^{3}H] corticosteroids are located in the hypothalamus, anterior pituitary, hippocampus and to a lesser extent in the amygdala and septum. However, since corticosteroids appear to affect certain hippocampal functions (see Chapter 11 Section 2), some of which are independent of the hypothalamic pituitary axis, the relative importance of the different target sites in relation to feedback mechanisms controlling ACTH secretion remain ill defined.

In the early 1940s it was discovered that the adrenocortical response to certain stressful stimuli could be blocked by previous administration of either adrenocortical extract or pure glucocorticosteroids. Subsequent demonstration of a similar inhibitory effect of glucocorticosteroids on pituitaries *in vitro* led to the view that the pituitary was the site of action of the negative corticoid feedback. In the 1960s, numerous investigators showed that injection or implantation of steroids in the hypothalamus and associated brain regions also effectively inhibited ACTH secretion implicating the nervous system in feedback inhibition. These experiments, however, suffer from the criticism directed against central administration of drugs, in that diffusion of steroid to distant sites, including the pituitary itself, may occur. The importance of the medial basal hypothalamus (MBH) or pituitary as targets for corticosteroid feedback was further implicated by elegant experiments in rats in which the MBH was surgically isolated from the rest of the brain (Halasz technique, see Chapter 2 Section 3(ii)). These animal preparations with 'hypothalamic islands' were devoid of the stress-induced and circadian rhythm of ACTH secretion of intact rats but their response to circulating corticosteroid was unaffected.

The physiological basis for corticosteroid regulation of ACTH secretion is not as straightforward as early experiments might have suggested. Experiments involving steroid treatment in 'non-stressed', adrenalectomised and intact rats generally show reciprocal relationships between blood levels of ACTH and corticosterone, as expected with a negative feedback system which operates at pituitary level; but the role of blood corticosteroids during stress-induced ACTH secretion is more complicated. In the early 1960s it was shown that there was no immediate correlation between plasma corticosterone concentrations and inhibition of the stress-induced activity of the hypothalamic-pituitary-adrenal axis. Subsequent experiments using controlled steroid infusion techniques showed that inhibition of the stress response occurred and then was absent for about a 2 hour delay after which inhibition returned without a further change in blood steroid levels. The concept of a 'rapid' followed by a 'delayed' feedback mechanism controlling the release of ACTH in response to stress is now generally accepted in the rat, and has recently been demonstrated in humans. Some evidence suggests the possibility that 'rapid' feedback involves inhibition of CRF and possibly ACTH release and operates through receptors sensitive to rates of increase in steroids. Whilst delayed feedback probably involves steroid-induced synthesis of a factor which eventually inhibits CRF synthesis, corticosteroids also appear partly to inhibit CRF-induced synthesis of ACTH. As a result of recent investigations various corticosteroids other than the main glucocorticoids (cortisol in man and corticosterone in rat) have been shown to inhibit the hypothalamic-pituitary-adrenal axis including the main mineralocorticoid, aldosterone. At the present time it is not clear how these relate to rapid and delayed feedback or to the various sites of the feedback action although differences have been observed. Further lines of evidence have suggested that oestrogenic (but not androgenic) gonadal steroids have a positive feedback action on the hypothalamic-pituitary-adrenal axis, but the physiological significance of minor glucocorticosteroids and oestrogens in the regulation of the hypothalamic-pituitary-adrenal axis requires further investigation. Finally, it was demonstrated as early as the 1950s, and more recently confirmed by *in vivo* and *in vitro* experiments, that the secretion of ACTH is inhibited via a 'short loop' feedback of the pituitary hormone itself acting on the hypothalamic CRF neurones. This mechanism may be important in times of severe stress. For further reading on corticosteroid feedback mechanisms see [6, 7].

5:6 Corticosteroid Receptors

Corticosteroids are only biologically active in the free form, although generally less than six per cent of the circulating corticosteroids are unbound. Most are bound to plasma proteins – 80 per cent to a specific cortisol binding globulin (transcortin) and approximately 14 per cent to albumin. In spite of its name, cortisol binding glubulin not only binds cortisol but also corticosterone and progesterone, but does not bind with aldosterone. An equilibrium exists between the bound and free circulating steroids and thus the high proportion of bound corticosteroids provides a large circulating reserve pool. The half life of corticosteroids in the plasma is in the order of 60 min. Although corticosteroids only interact with their receptors in the free form there is no indication of further metabolism prior to hormone–receptor interactions, as has been observed with testosterone for example (see Figure 8.6).

The distribution of corticosteroid binding cells in the brain and pituitary has been determined by injecting ^{3}H-labelled corticosteroids into adrenalectomised animals and subsequently subjecting the tissues to autoradiographic and biochemical procedures (see Chapter 2 Section 4(v)). As mentioned in the previous section labelled corticosteroids are bound in cells of the pituitary and hypothalamus and in other parts of the limbic system, notably the hippocampus. However, the intricacies of the receptors and potential feedback mechanisms are indicated by the finding that different labelled corticosteroids show different distributions of binding. For example corticosterone is bound in higher concentrations in the hippocampus compared with either the hypothalamus or amygdala whereas dexamethasone, a potent synthetic corticosteroid of widespread therapeutic value, shows higher affinity for the pituitary than for the brain. This may well explain the high potency of this drug to inhibit ACTH secretion.

The receptors for the glucocorticoid steroids appear to be similar to those for the gonadal steroids (see Chapter 8 Section 5) in that corticosteroids pass readily through the cell membrane, combine with a receptor in the cytosol and are translocated to a nuclear binding site where they induce changes in RNA and protein synthesis (see Chapter 1 Section 11) As for the gonadal steroids, it is not known precisely how corticosteroid receptor-interactions in the brain and pituitary are translated into alterations in the secretion rates of CRF and ACTH. It is possible that feedback mechanisms alter the turnover rates of certain neurotransmitters. Alternatively dexamethasone has a powerful inhibitory action on the electrical activity of hypothalamic neurones and this is due to a membrane effect. Thus it is possible that the feedback action of steroids on the CNS may also include direct alterations in the excitability of hypothalamic neurosecretory cells, or on cells that synapse with neurosecretory elements.

5:7 Neurotransmitters and CRF Regulation

As described in the previous sections, the synthesis and secretion of CRF is under rigorous control through a multitude of afferent neural pathways from different brain regions. These inputs involve different neurotransmitters which are either excitatory or inhibitory (see Figure 5.5). Many attempts to identify these transmitters have been undertaken although any direct approach is restricted by the lack of knowledge of the precise localisation of the CRF neurones.

In vivo techniques which have involved central administration or implantation of neurotransmitter drugs, with subsequent measurement of circulating ACTH levels, have shown that acetylcholine in the hypothalamus stimulates CRF secretion and its effect is blocked by the muscarinic cholinergic receptor antagonist, atropine. Similar experiments with the catecholamines, noradrenaline or dopamine have led

Figure 5.5: Control of ACTH Secretion: Possible Factors Involved. CRF – corticotrophin releasing factor; VP – vasopressin; DA – dopamine; NA – noradrenaline; 5-HT – 5-hydroxytryptamine; ACh – acetylcholine; RAS – reticular activating system; SON – supraoptic nuclei; SCN – suprachiasmatic nuclei; + – stimulatory; – – inhibitory.

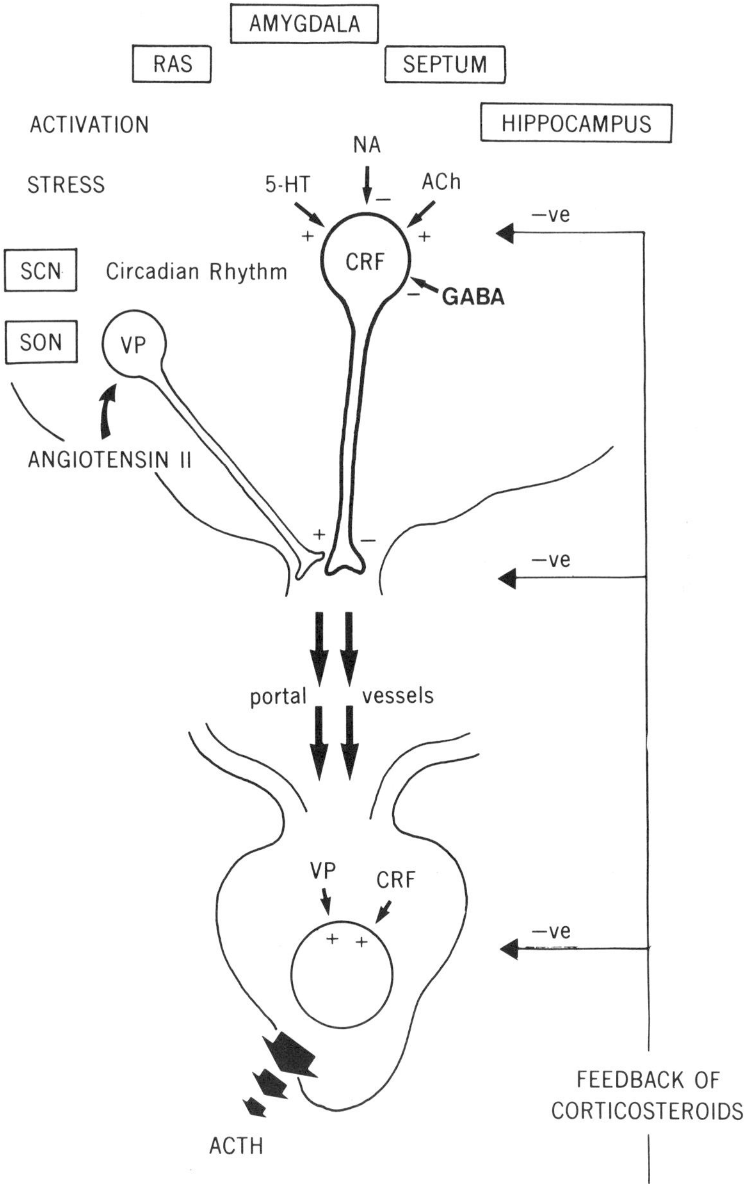

to the proposal that catecholaminergic innervation in the hypothalamus largely inhibits ACTH secretion. Implanting neurotransmitters or drugs in the extra-hypothalamic regions or in neural pathways converging on the hypothalamus have not produced consistent results, but noradrenaline and 5-hydroxytryptamine (5-HT) in regions of the limbic system including the amygdala, hippocampus and septal areas, alter ACTH secretion. These studies demonstrate that various converging neural pathways may involve different neurotransmitters and thus provide a basis for independent regulation of the circadian and stress-induced secretion of CRF and ACTH. Some evidence has been presented for example, that 5-HT is particularly implicated in the circadian changes in pituitary-adrenal function, since changes in plasma corticosterone and brain 5-HT levels in rats have been correlated. An alternative pharmacological approach which has confirmed the roles of the catecholamines involves the use of drugs, such as 6-hydroxydopamine, to induce the degeneration of specific amine pathways in brain. Such lesioned animals show a marked increase in ACTH secretion in response to stress. Further pharmacological experiments, using agonist or antagonist drugs which stimulate or block specific amine receptors respectively, have demonstrated that an important inhibitory influence on CRF secretion is due to noradrenaline mediated by the stimulation of α-adrenoreceptors, rather than β-adrenoreceptors.

In vitro studies on isolated hypothalamic tissue have in general confirmed *in vivo* investigations. Acetylcholine added to the isolated rat hypothalamus in low physiological concentrations stimulates the synthesis and release of CRF and similar doses of acetylcholine induce the release of CRF from isolated hypothalamic synaptosomes (nerve terminals), an effect which is totally inhibited by atropine. The action of acetylcholine on the intact hypothalamus, however, is only partially inhibited by atropine but completely blocked by the nicotinic cholinoreceptor antagonist, hexamethonium. These demonstrations may be explained by two types of stimulatory innervation of the 'CRF neurone' by acetylcholine – one which operates on the cell body and one at the level of the neurosecretory nerve terminal (see Figure 1.8). Inhibitory roles for both noradrenaline, dopamine and also the putative amino acid transmitter, γ-aminobutyric acid (GABA) on the release of CRF has also been confirmed by *in vitro* experiments whilst 5-HT stimulates CRF release from hypothalamic fragments. Thus experiments related to the putative neurotransmitters involved in the regulation of CRF secretion have provided potentially valuable information, although definitive evidence of their physiological importance requires the identification of the CRF neurones and of the particular synapses which relate to or directly impinge upon the neurosecretory cells. For further information see [7, 10].

5:8 Drugs Acting on the Hypothalamic Pituitary Adrenal Axis

A common stimulus for ACTH secretion is insulin induced hypoglycaemia (reduction of blood glucose caused by insulin injection). This is the basis for a test of hypothalamic-pituitary-adrenal integrity in cases of suspected dysfunction (see Chapter 12 Section 2). Alternative agents, which have been shown experimentally to stimulate CRF and thus ACTH secretion in animals, include histamine administration and inhalation of ether vapour. The opiate drugs, such as morphine and the natural opioid peptides, Met- and Leu-enkephalin have been shown to stimulate the hypothalamic-pituitary-adrenal axis by inducing CRF secretion directly but also have indirect effects operating through the CNS. However, prolonged treatment with morphine leads to an inhibition of pituitary-adrenocortical activity. There are a large number of other drugs which, when administered acutely, indirectly stimulate CRF secretion by acting on neurotransmitter systems in the CNS. These include reserpine (which depletes brain catechol- and indoleamines) and the neuroleptic drug, chlorpromazine (which has been shown to alter noradrenaline, 5-HT, acetylcholine and GABA neurotransmitter systems in brain).

A number of drugs abolish the secretion of CRF in response to stress and these have been particularly useful for the development of *in vivo* assays for CRF or ACTH. Often drugs which stimulate CRF and ACTH secretion acutely will when chronically administered, block subsequent stress-induced responses. In this category are reserpine, chlorpromazine and morphine. Administration of the latter two drugs following pentobarbitone anaesthesia have proved the most effective in inhibiting stress-induced hypothalamic-pituitary-axis activity. Alternatively the natural and synthetic corticosteroids may be used in supraphysiological amounts to bring about similar inhibition (see Section 5). For further information see [2].

References

1. Brodish, A. (1979) Control of ACTH Secretion by Corticotrophin-Releasing

Factor(s). *Vitamins and Hormones*, **37**: 111-53.
2. Buckingham, J. C. (1980) CRF. *Pharmacological Reviews*, **31**: 253-7.
3. Fink, G. (1981) Has CRF Finally been Found? *Nature*, **294**: 511-12.
4. Gann, D. S., Ward, D. G. & Carlson, D. E. (1978) Neural Control of ACTH: a Homeostatic Reflex. *Recent Progress in Hormone Research*, **34**: 357-400.
5. Hope, J. & Lowry, P. J. (1982) Pro-opiocortin: the ACTH/LPH Common Precursor Protein, in van Wimersma, T. B. Greidanns & L. Ress (eds), *ACTH and LPH in Health and Disease* (Karger-Basel).
6. Jones, M. T. *et al.* (eds) (1979) *Interaction Within the Brain - Pituitary - Adrenocortical System* (Academic Press, London).
7. Jones, M. T. (1978) Control of Corticotrophin (ACTH) Secretion, in S. L. Jeffcoate & J. S. M. Hutchinson (eds) *The Endocrine Hypothalamus* (Academic Press, New York), pp. 386-419.
8. Kopin, I. J. (1980) Catecholamines, Adrenal Hormones and Stress, in D. T. Krieger & J. C. Hughes (eds) *Neuroendocrinology – a Hospital Practice Book* (Sinauer Assoc. Inc., Sunderland, MA).
9. Schwyzer, R. (1977) ACTH: a Short Introductory Review. *Annals of the New York Academy of Science*, **297**: 3-25.
10. Vale, W., Rivier, J. & Rivier, C. (1980) Regulation of Hypophyseal Corticotrophic Cells, in J. L. Barker & T. G. Smith (eds) *The Role of Peptides in Neuronal Function* (Marcel Dekker Inc., New York), pp. 431-54.

6

CONTROL OF THYROID STIMULATING HORMONE

6:1 The Hypothalamic-Pituitary-Thyroid Axis

The synthesis and secretion of the thyroid hormones, thyroxine (T_4) and tri-iodothyronine (T_3) within the thyroid gland (see Figure 6.1) is controlled by thyroid stimulating hormone (TSH) secreted by the anterior pituitary gland. The thyroid hormones are unique, widely active hormones which play an important role both in growth and development and in energy balance by stimulating the rate of oxygen uptake and metabolism in most body tissues.

The thyroid gland synthesises and secretes T_4 but much of this is converted in blood and tissues into T_3 which may be the primary active hormone (see Section 6:6). However, details of the function of the thyroid hormones and of their production and synthesis are beyond the scope of this chapter and the interested reader is referred to [6, 10, 11]. It is now well established that TSH secretion and synthesis is under direct control of the hypothalamic thyrotrophin releasing hormone (TRH) and the feedback effects of circulating T_3 and T_4. This chapter summarises the complex neural control of TSH secretion and how TRH and the thyroid hormones interact, together with other factors, at the pituitary and in the hypothalamus to determine TSH release (see Figure 6.2). The effects of hypothalamic–pituitary dysfunction on the thyroid axis and diagnostic and therapeutic applications of TRH are covered in Chapter 12.

Figure 6.1: Thyroid Hormones

PRECURSORS

3-monoiodotyrosine

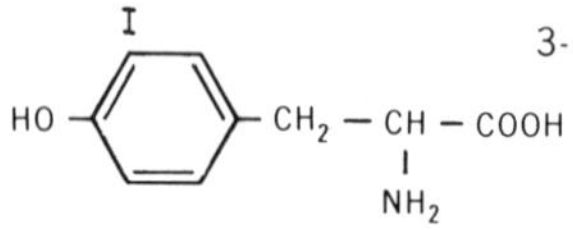

3,5-diiodotyrosine

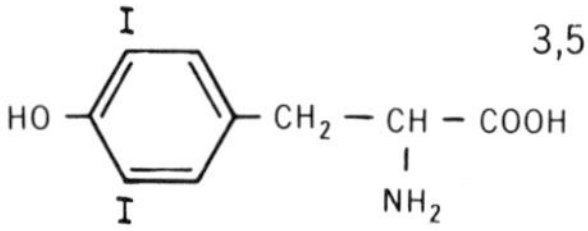

THYROXINE (T_4)

3,5,3′,5′-tetraiodothyronine

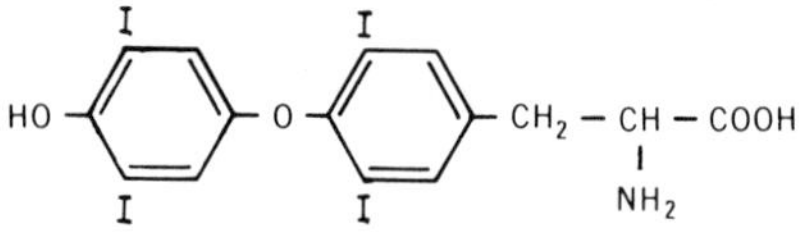

TRIIODOTHYRONINE (T_3)

3,5,3′-triiodothyronine

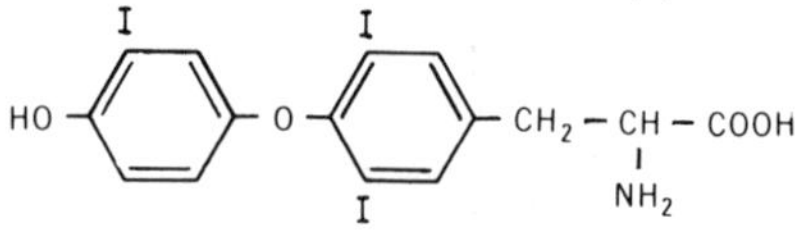

REVERSE (T_3)

3,3′,5′-triiodothyronine

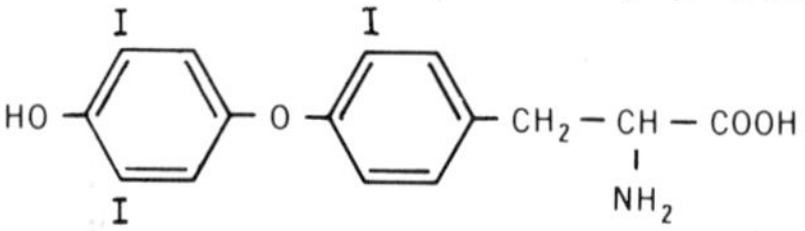

6:2 Thyroid Stimulating Hormone (TSH)

(i) *Structure of TSH*

TSH is a glycoprotein hormone secreted by a distinct group of basophilic cells in the anterior pituitary called thyrotrophs. The structure of TSH differs slightly between mammalian species but generally consists of two associated polypeptide subunits, α and β. The α unit

Figure 6.2: Control of TSH Secretion: Possible Factors Involved. TRH – thyrotrophin releasing hormone; SOM – somatostatin; DA – dopamine; NA – noradrenaline; 5HT – 5-hydroxytryptamine; HIS – histamine; TSH – thyroid stimulating hormone; GH – growth hormone; T_4 – thyroxine; T_3 – tri-iodothyronine; SCN – suprachiasmatic nucleus.

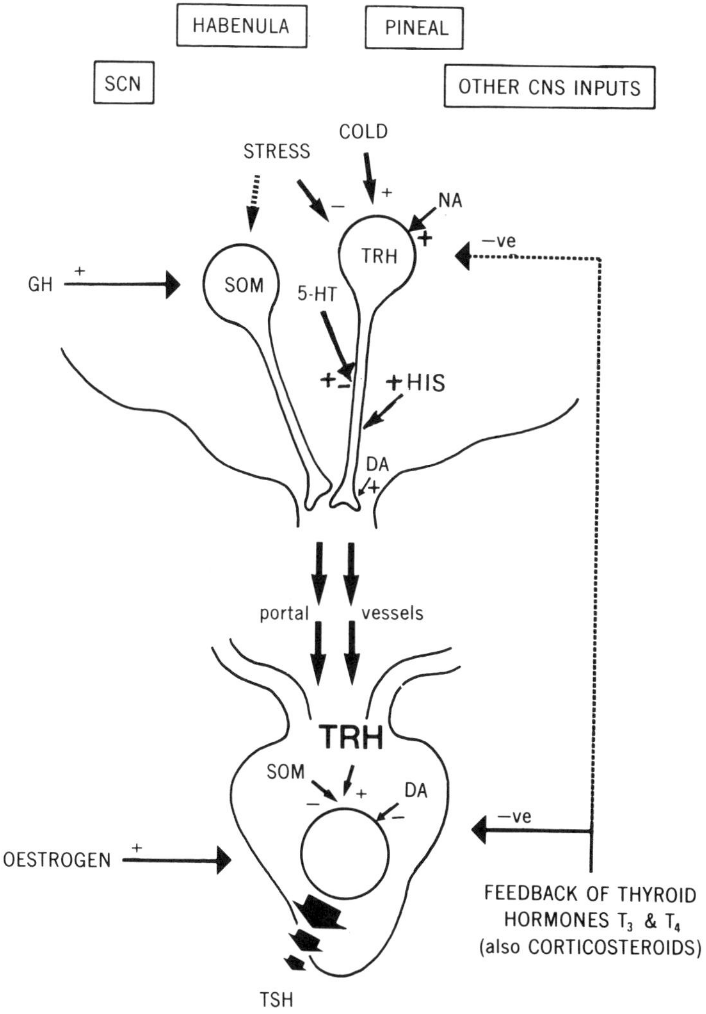

is identical with that of the other glycoprotein hormones, the gonadotrophins, (see Chapter 8 Section 2) suggesting a common evolutionary derivation of these hormones, while the β unit confers biological specificity on the hormone. Human TSH has a molecular weight of ~ 28,000 and the pituitary and circulating hormone appear identical. TSH is transported unbound in blood with a half life of approximately 60 minutes. For further information on the structure and function of glycoprotein hormones see [7].

(ii) *Measurement of TSH*

Radioimmunoassays for TSH and the gonadotrophins (see Chapter 2 Section 4(iii)) have been developed which are sufficiently sensitive to measure basal blood plasma levels in humans and experimental animals (~ 50 pg/ml) and the immunological as well as the biological specificity of the hormones resides in the small differences in amino acid content of their respective β subunits. There are also several bioassays for TSH which are based on the effects of this hormone on thyroid gland activity. Hence TSH concentrations can be assessed by measuring the secretion of T_3 and T_4 hormones utilising a radioimmunoassay, by measuring the rate of uptake of radioactively labelled iodine by the thyroid gland or even by histological examination of thyroid tissue since the amount of colloid present is inversely related to the rate of release of thyroid hormones.

The secretory pattern of TSH, like other pituitary hormones, is episodic with peaks occuring every 2-4 hours. Superimposed on this pulsatile secretion, a circadian periodicity has been demonstrated with maximum secretion rates occurring in the morning before awakening and minimum TSH release occuring in the late afternoon. Despite this there is no marked circadian variation in thyroid hormone levels in blood, suggesting that the pattern of TSH release is mediated by TRH secretion which itself is relatively independent of circulating levels of T_3 and T_4 (see Section 5). The action of TSH on the thyroid gland is to regulate the acquisition of iodine and also the various stages in the manufacture of the thyroid hormones from iodine and the amino acid tyrosine. T_4 (tetra-iodothyronine) is derived from two di-iodotyrosines and T_3 from a mono- and a di-iodotyrosine (see Figure 6.1). Also TSH stimulates the secretion of T_4 and T_3, and while T_3 constitutes only about 20 per cent of total hormone secreted by the thyroid gland it is thought to be the more biologically active thyroid hormone and that T_4 is converted into T_3 at specific target sites (see Section 6).

6:3 Thyrotrophin Releasing Hormone

(i) *Structure*

Thyrotrophin releasing hormone, which is released from hypothalamic nerve terminals in the median eminence and stimulates the secretion of TSH, was the very first hypothalamic hypophysiotrophic hormone to be isolated and chemically identified. Late in 1969, two American research groups under R. Guillemin and A. V. Schally independently identified an identical structure for sheep and pig TRH respectively (see Figure 6.3). A specific tripeptide with chemically blocked *N*- and *C*-terminal ends – pyro-Glu-His-Pro-NH_2 – which appears common to most species including man. The discovery of the structure of TRH culminated many years of effort by many laboratories to obtain direct evidence for the existence of hypothalamic releasing hormones, as predicted by G. W. Harris and others, and opened the doors for the avalanche of neuroendocrine research which was to follow.

(ii) *Localisation*

Localisation of TRH within the hypothalamus has been studied by radioimmunoassay of extracted tissues and by immunohistochemical methods both of which have depended on the ability to raise highly specific antibodies to this small peptide. Apart from the high concentrations of TRH-containing nerve terminals in the median eminence, immunoreactive TRH is found associated with cell bodies in the dorsomedial nucleus of the hypothalamus and the suprachiasmatic nucleus. Relatively high concentrations of TRH have also been measured in the ventromedial nucleus, the anterior hypothalamus and preoptic area, the periventricular region and the paraventricular nucleus (see Figure 1.2). Although the hypothalamus contains the highest concentration of TRH within the brain more than 80 per cent of TRH in the central nervous system (CNS) is found in areas other than the hypothalamus (see Chapter 10 Section 2). The brain stem, thalamus, septum and nucleus accumbens contain significant concentrations while the cerebral cortex and cerebellum contain relatively small amounts. The spinal cord is also rich in TRH with the highest concentrations in the ventral spinal cord, (anterior horn) and central canal regions. The distribution of TRH has primarily been investigated in the rat brain though recent data from human brain indicate a similar hypothalamic and extrahypothalamic distribution. TRH is also present in tissues outside the CNS, including the alimentary tract, pancreas, placenta and retina of the eye. For further information see [2].

Figure 6.3: Structure of TRH, its Metabolites and Some Peptidase-resistant analogues.

THYROTROPHIN RELEASING HORMONE

(a) TRH STRUCTURE :

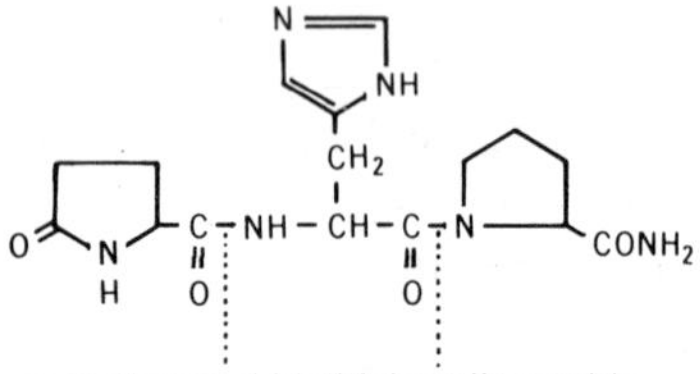

pyroglutamyl-histidyl-prolineamide
(pyroglu-his-proNH$_2$)

(b) TRH METABOLITES :

TRH-OH
(acid-TRH)

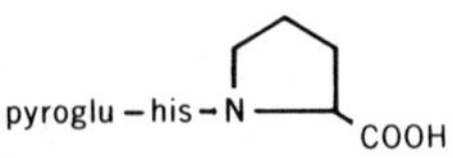

Cyclo-his-pro
(histidyl-proline-diketopiperazine)

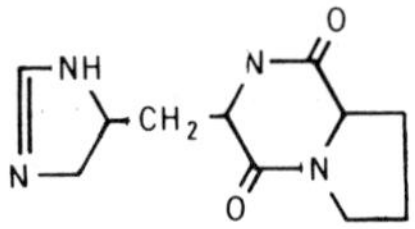

(c) TRH ANALOGUES :

1. N-terminal pyroglu:modification-orotyl-histidyl-prolineamide

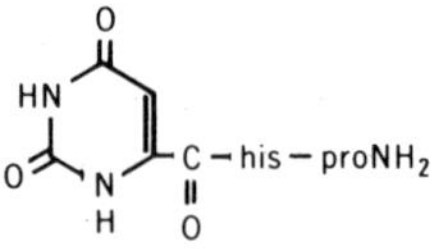

2. C-terminal proNH$_2$ modification:pyroglutamyl-histidyl-3,3′ dimethylprolineamide

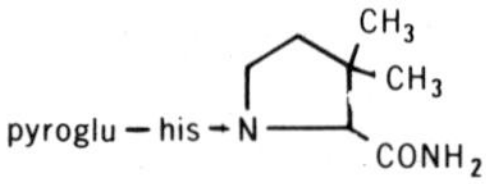

3. Histidyl modification:pyroglutamyl-3 methyl-histidyl-prolineamide

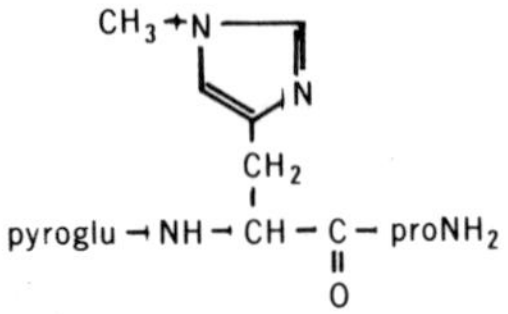

The precise role of extrahypothalamic TRH is not established but various lines of evidence point to a neurotransmitter or neuromodulator function although at the present time little is known of the TRH neuronal pathways (see Chapter 10 Section 3 and Chapter 11 Section 5). Nevertheless it appears that extrahypothalamic TRH is largely independent of hypothalamic TRH since both surgical isolation of the medial basal hypothalamus ('hypothalamic island') and lesions of the 'thyrotrophic' area deplete hypothalamic TRH without altering extrahypothalamic content. In the hypothalamus, as in other brain regions, TRH is present in both neurone cell bodies and nerve endings and is released from the terminals where, in common with other neurosecretory peptides, it is stored in granules prior to release. For further reading on extrahypothalamic TRH see Chapter 10 Section 2, Chapter 11 Section 7 and [4, 5, 8, 12].

(iii) *Synthesis and Release*

The mechanism of production or biosynthesis of TRH within the neurosecretory neurones is not understood. Some tenuous evidence was presented to suggest that enzymes were present to make TRH from its constituent three amino acids but this has not been accepted and, despite the lack of alternative evidence, the more generally held view is that TRH, similar to other neuropeptide hormones, is synthesised ribosomally as part of a larger polypeptide in the neurone cell bodies and is then processed from the precursor during transportation to the nerve terminals (see Chapter 3 Section 3). Preliminary reports of the isolation of a TRH precursor have been presented.

The hypothalamic-pituitary portal system is known to transport TRH to the pituitary, but evidence for an alternative route has recently been recognised whereby TRH, along with other peptide hormones, is secreted into the cerebro-spinal fluid (CSF) of the third ventricle and transported through ependymal tanycytes to the portal vessels (see Chapter 1 Section 8). The release of TRH has been investigated from hypothalamic fragments, brain slices and isolated nerve endings (synaptosomes) and in most cases it has been found that the release is dependent on the presence of calcium ions and can be stimulated by depolarising agents such as high molarity potassium and electrical stimulation.

(iv) *Action of TRH*

The action of TRH on pituitary thyrotroph cells leading to TSH secretion has been studied using purified isolated cells. TRH binds to high-affinity 'receptor' sites on the cell membranes and its effect on pituitary hormone secretion is mediated by changes in intracellular cAMP with a requirement for calcium ions – a mechanism common to many hormone actions (see Chapter 1 Section 10). TRH also stimulates the synthesis of TSH although the importance of this action in modulating pituitary responses is unknown. However, it is unlikely that TRH alone regulates TSH secretion. Experimental evidence has shown that the hypothalamic hormone, somatostatin, inhibits TSH secretion (Figure 6.2) and that neurotensin and possibly other unidentified factors promote the secretion of TSH. Furthermore, in addition to its effects on TSH secretion, TRH has also been shown to be a powerful stimulant of prolactin release by an action on the pituitary lactotrophs (see Chapter 7 Section 4). For several reasons the physiological significance of this action remains in doubt and, moreover, increases in the rate of prolactin release are not paralleled by increases in TSH secretion, nor are the respective peaks in the diurnal rhythms of TSH and prolactin release coincident. In contrast to these findings, a recent study has shown that peak levels of TSH (measured in peripheral blood) over a 24h period did not coincide with peak concentrations of TRH (measured in hypophyseal portal blood) although the preovulatory surge of prolactin in the proestrous rat did correlate with increased concentrations of TRH in hypophyseal portal blood. For further information see [1].

The presence of TRH outside the hypothalamus suggests that it has other actions apart form its hypophysiotrophic role. Electrophysiological experiments have shown that it can alter the firing frequency of both hypothalamic and extrahypothalamic neurones when applied iontophoretically to single cells (see Chapter 11 Section 5) and, although both excitatory and inhibitory effects have been reported, inhibition of excitability of neurones by TRH is the predominant response. Despite this, numerous behavioural studies have shown that either peripheral or central administration of TRH induces arousal and increased behavioural patterns such as grooming, preening, rearing and ambulation and TRH also counteracts the depressant effects of many drugs including neuroleptics, alcohol and anaesthetics. Also TRH has been found to stimulate a variety of sympathetic autonomic functions including heart rate, blood pressure and respiratory rate in both thyroidectomised and hypophyseotomised animals (see Chapter 11 Section 7) but the significance and underlying mechanisms of these effects are largely unknown. For further reading on extrahypothalamic actions of TRH see [1, 4, 5, 8, 12].

(v) *Degradation*

TRH is rapidly inactivated in blood, brain and other tissues by the action of several peptidase enzymes which either remove the *C*-terminal NH_2 alone (deamidation) or the *C*-terminal Pro-NH_2 or the *N*-terminal pyro-Glu. These enzymes have been partially purified and their activities characterised, and it has been suggested they may indirectly regulate TRH and TSH secretion although this remains to be convincingly demonstrated. Deamidation rapidly occurs in blood and tissues to form acid TRH (TRH-OH), an apparently inactive metabolite (see Figure 6.3). Another important metabolite is formed in brain and pituitary following removal of the *N*-terminal pyro-Glu. The initial dipeptide amide is unstable and readily cyclises to histidyl-proline diketopiperazine (cyclo His-Pro). Some of the reported biological effects of TRH may be mediated by cyclo His-Pro but it is clear that the conversion of TRH into this compound is not obligatory for its biological potency. In fact, it has been shown to antagonise certain TRH effects such as the TRH-stimulation of prolactin release and also certain TRH-induced CNS actions. A number of peptidase-resistant analogues of TRH have been synthesised (see Figure 6.3) with modifications to one or more amino acid residues. These are largely of research interest at the present time but such analogues may well provide useful drugs for the future (see Chapter 12 Section 3). Certain analogues are particularly potent on TSH and prolactin secretion (e.g. 3-methyl His TRH) whilst most others are more potent in brain. For further reading on TRH analogues see [1, 4].

6:4 Neural Control of TRH and TSH Secretion

(i) *Neural Factors – Cold and Stress*

Acute exposure to cold is a potent stimulus for TSH secretion. This TSH response to acute cold is rapid in onset, occurring within 10 minutes, and is accompanied by increased thyroid activity and changes in TRH content in certain regions of the hypothalamus. The effect can be abolished by lesioning the hypothalamus and conclusive evidence for the involvement of TRH in mediating this response has been provided by studies in which administration of antiserum to TRH inhibits the response to cold. Despite this reflex response to acute cold, the precise role and importance of the hypothalamic-pituitary-thyroid axis in the adaptation to prolonged or chronic exposure to cold is not clear. Other factors such as blood plasma catecholamines are active in thermogenic mechanisms and the thyroid hormones alter both the secretion and action of these amines.

In laboratory animals various stressful stimuli (with the exception of exposure to cold) such as physical restraint, electric shock or placing in a novel environment all tend to reduce TSH levels in blood suggesting inhibition of TSH secretion. The role of TRH in this stress response is uncertain. However, since the stress induced reduction in both growth hormone (GH) as well as TSH secretion in rats can be inhibited by administration of an antiserum to somatostatin, and also TRH induced TSH secretion is known to be inhibited by somatostatin, it is possible that the inhibitory effects of stress on GH and TSH secretion may both be mediated by somatostatin. The physiological responses to stress are very complex and vary depending on the nature of the stress, be it physical or emotional, acute or chronic (see also Chapter 5 Section 4). Although changes in the hypothalamic-pituitary-thyroid axis of the rat in response to a particular stressful stimulus may be clear cut, this is by no means true in primates (including man) and many factors in addition to the hypothalamic-pituitary axis may contribute to this. For example there may be changes in either the carrier binding proteins of the thyroid hormones or the peripheral metabolism of the hormones. Furthermore, recent biochemical studies have shown that T_4 may not only be metabolised to T_3 but also to reverse T_3 (rT_3) (see Figure 6.1). This is an inactive form of the hormone and it would appear that more rT_3 is synthesised during stress.

(ii) *Hypothalamic Control*

Surgical disconnection of the pituitary from the hypothalamus in mammals results in decreased TSH secretion and reduced thyroid hormone production (hypothyroidism). However, relatively normal TSH levels are observed in rats where the connection between pituitary and hypothalamus remains intact but the hypothalamus is disconnected from the rest of the brain (hypothalamic islands). This indicates that the so-called 'hypophysiotrophic area' of the hypothalamus is essential for normal maintenance of TSH secretion. Electrical stimulation of the anterior hypothalamus, using implanted electrodes in either the preoptic area, the paraventricular, dorsomedial or ventromedial nuclei, induces TSH release with a time course similar to that observed by administration of TRH. Such studies are in accordance with the immunohistochemical observations on the localisation of TRH-containing neurones in the hypothalamus. It is now generally agreed that the thyrotrophic area is situated in the anterior hypothalamic area although several studies have been undertaken in an attempt to delineate more precisely hypothalamic regions involved in the control TSH secretion. A recent study in which small surgical lesions were used, as opposed to electrolytic lesions, suggests that the paraventricular and periventricular nuclei are of particular importance in maintaining TSH secretion while the medial preoptic area tonically inhibits TSH release. Since most of the somatostatin-containing nerve cell bodies reside in the preoptic area and somatostatin is known to inhibit TRH-induced release of TSH, it is suggested that the tonic inhibition of TSH secretion exerted by the preoptic area may be mediated via somatostatin secretion.

(iii) *Extrahypothalamic Control*

The mediobasal hypothalamus is capable of maintaining basal TSH secretion and the circadian rhythms, but stress-induced inhibition and cold-induced release of TSH is probably controlled by extrahypothalamic brain regions. Among the extrahypothalamic structures, the habenula and pineal have received the most attention and both seem to have an inhibitory action on thyroid function. Extrapyramidal and limbic structures have also been implicated in regulating TSH secretion but relatively few studies concerning extrahypothalamic control of TSH have been reported. The wide distribution of TRH in extrahypothalamic brain areas suggested that these areas might have a role in TRH–TSH interactions but experimental evidence is lacking and TRH content in brain areas outside of the hypothalamus do not alter in parallel with thyroid hormone status. It is evident that studies concerning the extrahypothalamic control of TSH are few, fragmentary and inconclusive and the role of other brain regions remains speculative at the present time.

(iv) *Neurotransmitter Regulation*

The effects of cold on TSH release can be inhibited by drugs which deplete or block catecholamine neurotransmitters or by neutralisation of endogenous TRH by a TRH-antiserum, suggesting that cold stimulated TSH secretion is mediated by catecholamine synapses regulating neurosecretory activity. Other studies have also shown that noradrenaline may have a stimulatory role on TSH secretion by facilitating the release of TRH. Dopamine may exert inhibitory effects directly at the pituitary although a stimulatory action on TRH secretion has also been demonstrated particularly in *in vitro* studies using either median eminence fragments or isolated nerve endings (synaptosomes). Thus dopaminergic innervation of the median eminence may regulate TRH secretion. Contradictory results have been obtained from studies on the role of 5-hydroxytryptamine (5-HT) though the consensus is for an inhibitory control by 5-HT on the hypothalamic secretion of TRH. Furthermore, there is evidence from animal studies and from the use of isolated tissue slices that histamine (which appears to have a neurotransmitter function in the hypothalamus) may stimulate TRH secretion. In summary, however, the role of neurotransmitters in the control of TSH release is controversial, contrasting the evidence pertaining to neurotransmitter function in the hypothalamic regulation of other pituitary hormones. For further reading see [1, 9].

6:5 Thyroid Hormone Feedback

Removal of the thyroid glands (thyroidectomy) causes an increase in plasma TSH levels due to the loss of negative feedback by thyroid hormones, although even greater elevations of plasma TSH occur in thyroidectomised animals bearing hypothalamic lesions. Lesions of the anterior hypothalamus result in the complete disappearance of TRH from the median eminence and decreased TSH secretion, but acute withdrawal of T_3 and T_4 can increase plasma TSH levels despite the apparent absence of a hypothalamic influence. Conversely TRH stimulated TSH secretion can be progressively blocked by increasing doses of T_3 and T_4 while very low doses of T_4 are only effective in suppressing TSH secretion in thyroidectomised hypothalamic-lesioned rats. Observations of this kind clearly indicate that circulating levels of thyroid hormones exert a negative feedback action on the pituitary (Figure 6.2). Although several studies have been undertaken to determine a possible feedback action of thyroid hormones on the TRH neurosecretory cells in the hypothalamus, the results have been inconsistent with both positive and negative effects being reported and discrepancies between data obtained from bioassay or radioimmunoassay of hormone concentrations. Overall, the indication is that there is competition between TRH and thyroid hormones at the pituitary and it is thought that the rate of TRH secretion determines the sensitivity of the pituitary to the negative feedback effects of T_3 and T_4. These hormones are probably the primary modulators of TSH secretion at the level of the pituitary although hypothalamic somatostatin is also clearly involved. Furthermore, oestrogens can affect the sensitivity of the pituitary to TRH and it is thought that these steroids enhance TRH responses by increasing pituitary receptors for TRH. For further reading on the feedback regulation of TSH secretion by thyroid hormones see [3].

6:6 Thyroid Hormone Receptors

The thyroid hormones, T_3 and T_4 are transported in the blood bound preferentially to a thyroxine-binding globulin or to a pre-albumin and, as for circulating steroid hormones, it is only the small proportion of thyroid hormones circulating in a free form that is active and able to bind with receptors. The ubiquity of target tissues for the thyroid hormones, which have been demonstrated by binding studies with radioactively labelled hormones, support the evidence that the thyroid hormones stimulate the rate of oxygen uptake and metabolism in most body tissues. Only the spleen, lung, testes and brain show a negligible binding capacity for these hormones, although neonatal brain strongly binds labelled T_3 – a finding which is consistent with the evidence that the thyroid hormones are critical for the normal growth and development of the brain. In the adult brain however, thyroid hormones are reported to influence neurotransmitter function and behaviour (see [11]).

Again like the steroid hormones, thyroid hormones enter the cells of the target tissues and bind to the nucleus (see Chapter 1 Section 11), although, unlike the steroid hormone receptors, the existence of a cytoplasmic receptor and translocation to the nucleus is not in evidence. Instead thyroid hormones appear to interact directly with nuclear receptors which are bound to a non-histone chromatin protein. T_3 receptors are capable of binding to DNA but the nature of the DNA protein interaction is not fully understood. More recently, convincing evidence has been presented for the existence of mitochondrial receptors for thyroid hormones, and these may be responsible for the increased O_2 uptake and metabolism (calorigenic effect) induced by T_3 and T_4. The nuclear receptors have a high affinity for T_3 (kd – 10^{-10}M) but a much lower affinity for T_4 and it is generally considered that T_3 is the active thyroid hormone in target tissues. T_4 is readily deiodinated to T_3 both in the plasma and in the pituitary gland and it has been estimated that about 70 per cent of the bound T_3 is derived from an intrapituitary conversion of T_4 into T_3. The inhibitory effect of T_3 and T_4 on the pituitary can be blocked by protein synthesis inhibitors such as puromycin and cycloheximide and it is considered that T_3 and T_4 induce the synthesis of an inhibitory protein through a mechanism requiring DNA-dependent RNA synthesis. The question remains uncertain whether T_4 *per se* can regulate TSH secretion or must be totally converted into T_3 to be an effective stimulus on the pituitary. See [3] for further details.

6:7 Other Hormones and Drugs Affecting TSH Secretion

Adrenal corticosteroids can inhibit the secretion of TSH by acting on the pituitary thyrotrophs and also by reducing TRH secretion. In addition these steroids have been shown to inhibit the conversion of T_4 into T_3. Oestrogens may also influence the release of TSH, since the TSH response following TRH stimulation is greater in women than in men and, in the male, oestrogen treatment enhances pituitary responsiveness to TRH. Studies in the rat have correlated this effect with an increased number of receptors on both the thyrotrophs and lactotrophs following oestrogen treatment, so that the TRH induced release of both TRH and prolactin are enhanced. Certain evidence has led to speculation on the existence of a short-loop positive feedback of growth hormone (GH) on somatostatin secretion (see Chapter 7 Section 3(iii)). As pointed out above, somatostatin also inhibits TRH-induced TSH release, and thus growth hormone may also indirectly inhibit TSH secretion (see Figure 6.2). This may be the basis of the demonstration that growth hormone therapy, possibly via somatostatin release, causes hypothyroidism; although alternative studies have not supported this idea.

A large number of drugs used in medical practice are known to alter TSH secretion, usually indirectly by altering the production, transport in blood, metabolism or action of the thyroid hormones, T_3 and T_4. Antithyroid drugs (goitrogens) interfere, in one way or another, with the biosynthesis of thyroid hormone and the resulting fall in blood T_3 or T_4 can lead to augmented TSH secretion (see Chapter 12 Section 2). However, certain drugs appear to act directly on the hypothalamic control of TSH secretion. For example morphine, pentobarbital and reserpine have each been shown to cause a decrease in TSH secretion in this way (see [1, 9]).

References

1. Griffiths, E. C. & Bennett, G. W. (eds) (1982) *Thyrotrophin Releasing Hormone (TRH)* (Raven Press, New York).

2. Jackson, I. M. D. (1982) Thyrotrophin-Releasing Hormone. *The New England Journal of Medicine*, **306**: 145-55.
3. Larsen, P. R. (1982) Thyroid-Pituitary Interaction: Feedback Regulation of thyrotrophin secretion by thyroid hormones. *The New England Journal of Medicine*, **306:** 23-32.
4. Metcalf, G. & Dettmar, P. W. (1981) Is Thyrotrophin Releasing Hormone an Endogenous Ergotropic Substance in the Brain? *Lancet*, i; 586-9.
5. Morley, J. E. (1979) Extrahypothalamic TRH – its distribution and its functions. *Life Sciences*, **25**; 1539-50.
6. Oppenheimer, J. H. (1979) Thyroid Hormone Action at the Cellular Level. *Science*, **203**: 971-9.
7. Pierce, J. G. & Parson, T. F. (1981) Glycoprotein Hormones: Structure and Function. *Annual Review of Biochemistry*, **50**: 465-95.
8. Prange, A. J. & Utiger, R. D. (1981) What Does Brain Thyrotrophin-Releasing Hormone do? *The New England Journal of Medicine*, **305**: 1089-90.
9. Reichlin, S., Martin, J. B. & Jackson, I. M. D. (1978) Regulation of Thyroid Stimulating Hormone (TSH) Secretion, in S. L. Jeffcoate, J. S. M. Hutchinson (eds), *The Endocrine Hypothalamus* (Academic Press, New York), pp. 230-69.
10. Schimmel, M. & Utiger, R. D. (1977) Thyroidal and Peripheral Production of Thyroid Hormone: Review of Recent Findings and their Clinical Implications. *Annals of Internal Medicine*, 87: 760-8.
11. Singhal, R. L. & Rastogi, R. B. (1981) Thyroid Hormone in the Regulation of Neurotransmitter Function and Behaviour, in P. D. Halina & R. L. Singhal (eds) *Neuroendocrine Regulation and Altered Behaviour* (Croom Helm, London), pp. 204-21.
12. Yarborough, G. G. (1979) On the neuropharmacology of thyrotrophin releasing hormone (TRH). *Progress in Neurobiology*, **12**: 291-312.

7

CONTROL OF GROWTH HORMONE (SOMATOTROPHIN) AND PROLACTIN SECRETION

7:1 Growth Hormone (GH) and Prolactin (i) Structure (ii) Receptors for GH and prolactin (iii) Measurement of hormones (iv) Actions of growth hormone and prolactin **7:2** Hypothalamic Control of Growth Hormone Secretion – Growth Hormone Releasing Factor and Somatostatin (i) GH releasing factor (ii) Somatostatin – structure, synthesis and degradation (iii) Localisation and action of somatostatin **7:3** Neural Control of GH Secretion (i) Episodic GH secretion (ii) Physiological factors stimulating GH secretion (iii) Feedback regulation of GH release (iv) Effects of Neurotransmitters, neuropeptides and hormones on GH release **7:4** Hypothalamic Control of Prolactin Secretion – Prolactin Inhibiting and Releasing Factors (i) Dopamine as the prolactin inhibiting factor (PIF) (ii) Prolactin releasing factors **7:5** Neural Control of Prolactin Secretion (i) Rhythms of prolactin release (ii) Prolactin release in response to suckling and stress (iii) Hormonal modulation of prolactin secretion (iv) Neurotransmitters and neuropeptides in the hypothalamic control of prolactin secretion

This chapter is concerned with the neural mechanisms which control the secretion of growth hormone (somatotrophin) and prolactin (luteotrophic hormone) and, though they have different biological functions in the body, there are several reasons for discussing the two hormones together. Firstly they are structurally related hormones, secondly they do not have specific target endocrine glands, as do corticotrophin, thyroid stimulating hormone and the gonadotrophins, and thirdly both hormones appear to be controlled by a dual hypothalamic mechanism involving both releasing and release-inhibiting factors or hormones.

7:1 Growth Hormone and Prolactin

(i) *Structure*

Growth hormone (GH) and prolactin are structurally related single-chain polypeptide hormones with molecular weights in the order of 20,000-25,000. They are homologous proteins and their primary structures are identical at approximately 25 per cent of all their amino acids though this figure varies between species. They also bear a marked structural resemblance to the placental hormone, chorionic somatotrophin or placental lactogen, which has growth promoting activity. However, there is considerable species variation in both their structures and biological actions. The fact that GH from lower mammals is inactive in man, but conversely primate GH has a high affinity for the receptors in lower mammals, suggests that the hormones may have evolved from a single hormone and evolution has conferred specificity on the hormones and/or their receptors. The primary structure of GH and prolactin has been elucidated for several mammalian species; human GH contains 191 amino acids with two disulphide bridges and human prolactin has 198 amino acids with three disulphide bridges (see Figure 7.1). Comparison of human GH with human prolactin shows that only 16 per cent of the amino acids are in identical sequence positions although human prolactin is very similar to ovine prolactin with 73 per cent of the residues being in the same sequence. The linear structure of human GH and ovine prolactin with the positions of the disulphide bridges are shown diagramatically in Figure 7.1 which also shows regions of internal homology in the sequence of the amino acids. Their stable three-dimensional (tertiary) structure is unknown although it is likely that this is important in determining their interaction with receptors. However, up to 20 per cent of the hormone can be removed by enzymic degradation without loss of activity, indicating that certain peptide sequences in the molecule are not essential for biological potency; even intact disulphide bridges do not seem essential. The hormones are stored in and secreted by acidophil cells in the anterior pituitary and GH is the most abundant hormone in the gland (estimated ~ 8 per cent gland weight). The somatotrophs or GH cells (which, in the human, are located predominantly in the posterolateral regions of the pars distalis) contain round dense secretory granules about 300-350 nm in diameter (see Plate 2, Chapter 2). The mammotrophs (lactotrophs) or prolactin cells are rather scarce, except during pregnancy and lactation, and are diffusely distributed throughout the pars distalis. Their mature secretory granules are large (600-700 nm) and irregularly shaped making the prolactin cell easy to identify.

(ii) *Receptors for GH and Prolactin*

Binding studies with ^{125}I-labelled GH and prolactin suggest that cellular receptors are located on the cell membrane and are rather widely distributed throughout the body. Both hormones bind to the liver and kidney in high concentrations but this may be related to their metabolism

Figure 7.1: (a) Schematic Structures of Human Growth Hormone and Ovine Prolactin Showing Positions of the Disulphide Bridges. (b) Diagrammatic representation of the homologous regions in the sequence of amino acids, indicated by cross-hatched areas.

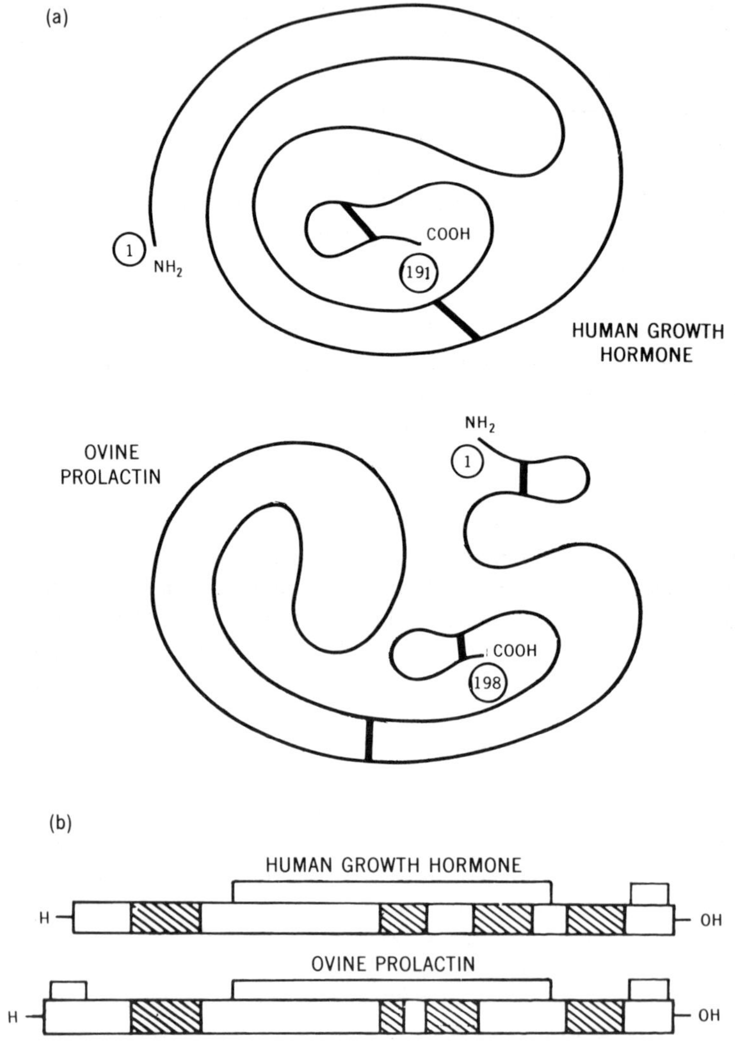

rather than with mediating their biological actions. Prolactin binding also occurs in the testis, prostate, ovary and mammary gland (particularly during lactation), whereas labelled GH is recovered in relatively high concentrations from the spleen, pancreas, intestine, adrenal cortex and both heart and skeletal muscle.

Little can be said about the immediate cellular reactions that occur after the binding of these hormones with their receptors although a few GH effects appear to be mediated via cyclic AMP (see Section 1:10). Furthermore, it may be that at some target tissues these hormones may enter the cell in order to bring about their biological actions since there is evidence that the prolactin–receptor complex may be taken up by liver cells (internalisation – see Chapter 1 Section 10) and combine with the Golgi apparatus. However, whether the process of internalisation is important for the initiation of biological response or represents a stage in the degradation of the hormone remains to be established. Cellular events following the binding of prolactin and GH to receptors in their target cells include alterations of transport of amino acids and other metabolites, induction of specific proteins and nucleic acid synthesis.

(iii) *Measurement of GH and Prolactin*

The most commonly used bioassay for GH has been the tibia test. It entails the administration of the hormone sample to female rats or mice, which have been hypophysectomised to remove the animals' endogenous GH, and the determination of the increased width of the proximal epiphysial cartilage of the tibia some 2–3 weeks later. The tibia test is sufficiently sensitive to measure high blood levels of GH as occur in gigantism and acromegaly or levels which occur in response to insulin-induced hypoglycaemia (see Chapter 12 Section 2) but the test cannot detect normal resting blood levels (10 mU/l). Alternative bioassays are based on incremental changes in body weight or increased nitrogen retention, but measurement of basal levels and subtle changes in secretion requires the application of a sensitive and specific radioimmunoassay. The most common bioassay for prolactin is the stimulated formation of crop 'milk' in pigeon crop sacs. Another bioassay to determine the presence of prolactin activity has been the stimulation of the drive to migrate to water in certain amphibian species – 'the eft water drive' – but as for GH these assays were unable to detect low circulating levels of prolactin and thus their experimental application was of restricted value.

It was not until the detailed structures of human and primate GH and prolactin were more clearly elucidated in the early 1970s that research into neuroendocrine mechanisms regulating their secretion really became established. The main reason for the more recent progress was the development of radioimmunoassays for the measurement of these hormones (see Chapter 2 Section 4(iii)). It should be noted that for GH in particular, there have been a number of conflicting results between bioassay and radioimmunoassay determinations of blood levels. Reasons for this relate to rapid turnover or metabolism of the hormone and the production of mediators in certain biological actions.

(iv) *Actions of Growth Hormone and Prolactin*

The most obvious action of growth hormone (GH) is its marked growth promoting effect on bone, cartilage and other tissues. This largely occurs as a result of the stimulation of protein and collagen synthesis and the uptake and utilisation of circulating amino acids by the tissues. GH is essential for maintaining linear bone growth, but it is now known that this is not a direct action of GH but is mediated by the production of growth factors in blood. The prime factors are peptide hormones (mol. wt. $\sim$ 4,000) called somatomedins which are synthesised in the liver and kidneys. Further details regarding the production and actions of somatomedins can be obtained in [11, 13]. Although GH is essential for the growth of most tissues it is apparently less critical for normal brain development than the thyroid hormones. Young hypophysectomised animals, in which the source of GH and other pituitary trophic hormones has been removed, stop growing within a few weeks and die at an early age. One reason for this is due to brain damage, since the brain continues to develop within a cranium or skull which has ceased to grow. In addition to promotion of body growth and a synergistic action on the trophic effects of other hormones, GH has important roles in the metabolism of proteins, carbohydrates and fats. The general protein anabolic effects result in maintaining a positive balance of nitrogen and phosphorus in the body; responses which are also encouraged by insulin. However GH, together with the adrenal corticosteroid hormones (see Figure 5.1) antagonise the effects of insulin on glucose uptake and utilisation by muscle and other tissues, and also breaks down liver glycogen stores to increase blood levels of glucose. Similarly GH stimulates lipolysis and the mobilisation of fatty acids, but the possible role of somatomedins in the metabolic actions of GH is currently unsettled. Not surprisingly excess GH shows certain prolactin-like (lactogenic) actions.

Prolactin is a remarkably versatile hormone found in all vertebrates and numerous and diverse functions for this hormone have evolved in addition to a general growth promoting activity. In fact over eight-five different functions for prolactin have been described ranging from osmoregulation in fishes, through water migration in amphibia and maternal behaviour in birds, to milk production in mammals. Its role in lactation is probably the most well-defined action in mammals, but it should be remembered that numerous other hormones, such as oestrogen, progesterone and corticosteroids, are also required for development of mammary glands and milk production. During pregnancy prolactin secretion increases gradually and the lactational effects of this hormone are thought to be inhibited by the high circulating levels of oestrogen. Oestrogen levels fall at the time of parturition thus allowing prolactin to exert its full effects on lactation. Interestingly, the amniotic fluid also contains high concentrations of prolactin and it has been suggested that it may play a role in salt and water balance of the foetus, reflecting a primitive osmoregulatory function of the hormone as found in fishes. Other effects of prolactin have been described including a synergistic action with noradrenaline on contraction of vascular smooth muscle, stimulation of corticosteroid production, and effects on gonadal function. Further details on the possible effects of prolactin on gonadotrophin secretion are given in Section 5(iii) and the action of prolactin on gonadal receptors and trophic effects on the corpus luteum are described in Chapter 8 Sections 2 and 1 respectively. For further reading on prolactin and its diverse functions see [1, 2, 7, 8].

7:2 Hypothalamic Control of Growth Hormone Secretion – Growth Hormone Releasing Factor and Somatostatin

(i) *GH Releasing Factor*

There are at least two hypothalamic factors or hormones which regulate the secretion of GH following their release into the hypophyseal portal vessels, one stimulatory and the other inhibitory. The predominant hypothalamic influence on GH secretion is stimulatory in nature which contrasts with the hypothalamic control of prolactin release. The growth hormone releasing factor (GH-RF) has not, to date, been identified or characterised although a variety of experimental data leave little doubt of its existence. Crude and semi-purified extracts of hypothalamus effectively stimulate GH release in animals *in vivo* and from isolated pituitary tissue *in vitro*. Although various known factors in the hypothalamus, including peptides such as vasopressin and TRH cause GH release, none of these are considered to be the specific GH-RF. Interestingly, a hypothalamic decapeptide with appropriate bioactivity was isolated in 1971 by A. V. Schally and colleagues but this did not release immunoreactive GH in a number of species and was subsequently identified as a fragment of the blood protein, haemoglobin. The precise localisation of neurones which secrete GH-RF has not been determined since immunological techniques are not applicable to peptides of unknown structure. However, stimulation and lesion studies have shown that the cell bodies which may contain a GH-RF are likely to be in the ventromedial nucleus (VMN) of the hypothalamus.

(ii) *Somatostatin – Structure, Synthesis and Degradation*

During the search for GH-RF, some side fractions of purified hypothalamic extract were found to inhibit GH release and in 1968 a peptide in hypothalamic extracts with GH release-inhibiting activity was first described. In 1973 Guillemin and his collaborators finally characterised, sequenced and synthesised this peptide, which was subsequently named somatostatin.

Somatostatin is a cyclic tetradecapeptide with one disulphide bond between Cys 3 and Cys 14 although the reduced linear form of the peptide is equally active (see Figure 7.2). Similar to the synthesis of ACTH and related peptides from a pro-opiocortin precursor (see Chapter 5 Section 2(ii)) somatostatin also appears to be derived from part of a larger pre-prohormone containing in excess of 90 amino acids (see [14]), but, in contrast with pro-opiocortin no other amino acid sequences in this precursor molecule, apart from the *C*-terminal amino acid sequence of somatostatin, appear to resemble other known biological peptides. The half life of somatostatin in blood is very short (~ 4 min) and the points at which the hormone is cleaved by the

Figure 7.2: Structure of Somatostatin.

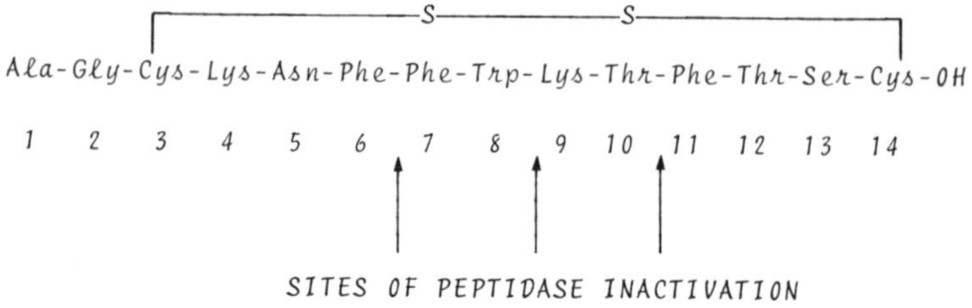

action of certain peptidase enzymes is indicated in Figure 7.2. Studies on inactivation showed a preferential release of internal residues involving the Trp^8-Lys^9 linkage and to a lesser extent bonds adjacent to the Phe-residues. In addition a slow release of Ala^1 has been observed but since neither Ala^1 nor Gly^2 appear essential for biological activity it is unlikely that this is a primary mechanism for inactivation. More than one enzyme system is available for the inactivation of somatostatin and peptidases both in serum and intracellularly in the cytoplasm and lysosomes of central and peripheral cells have been identified. Thus regulation of the biological activity of the peptide will be dependent on its localisation with respect to the degradation system involved.

(iii) *Localisation and Action of Somatostatin*

Studies on the distribution of somatostatin by radioimmunoassay and immunohistochemistry have indicated that this peptide is widely distributed throughout the CNS. In the brain, the most abundant somatostatin-containing nerve endings are located in the median eminence. Their neurone cell bodies are found in various parts of the hypothalamus but particularly in the anterior periventricular regions including the preoptic area. These neurones are probably the neurosecretory cells which exert the inhibitory control of GH secretion. There is a heavy representation of immunoreactive somatostatin in the limbic system (amygdala, hippocampus, nucleus accumbens and the olfactory tubercle) and numerous somatostatin-containing fibres have been located in the caudate nucleus, cerebral cortex and the brain stem. In the spinal cord, fine afferent nerve fibres which terminate in the dorsal horn contain somatostatin and this peptide is also present in some autonomic ganglia and is one of several peptides found in the retina (see Table 10.1). The widespread distribution of somatostatin in the central nervous system certainly implicates a neurotransmitter or neuromodulator function and the effects of this peptide on strychnine-induced convulsions, the duration of barbiturate anaesthesia and certain behavioural patterns, such as feeding and drinking, support this idea. In the spinal cord it is thought to be an inhibitory neurotransmitter in afferent sensory pathways and in the brain it has been shown to inhibit the firing rate of neurones (see Chapter 11 Section 5).

Somatostatin released from nerve terminals in the median eminence is transported to the anterior pituitary via the hypophyseal portal vessels or via the tanycytes after its secretion into the cerebrospinal fluid (see Chapter 1 Section 8). The actions of somatostatin on the pituitary gland are to inhibit GH, and also TSH, release and the physiological significance of these effects is indicated by the ability of an antiserum to the hormone to elevate circulating levels of GH and TSH. The mechanism of action of somatostatin on the somatotrophs has not been clarified. It has been suggested that it may inhibit the accumulation of cyclic AMP in anterior pituitary cells although other studies implicate that somatostatin can exert its effects beyond the action of common cell mediators such as cyclic AMP and prostaglandins (see Chapter 1 Section 10). The effects of somatostatin on GH secretion are rapid and usually of short duration and apart from the involvement of somatostatin in the regulation of episodic GH release it can also inhibit increases of GH release stimulated by insulin-induced hypoglycaemia or sleep. The inhibition of TSH release by somatostatin appears to be exerted by antagonising the action of TRH on the pituitary thyrotrophs (see Chapter 6 Section 7). Interestingly somatostatin has no apparent effect on the TRH stimulation of prolactin secretion, suggesting that its effects on TSH secretion are specific. In normal circumstances the basal or induced release of corticotrophin and the gonadotrophins is also unaffected by somatostatin administration.

The finding that somatostatin would inhibit the release of insulin and glucagon from the endocrine pancreas led to the observation that immunoreactive somatostatin is present in the delta cells of the pancreas and subsequently it was shown that somatostatin was widely distributed throughout the gastrointestinal tract. Numerous studies have been carried out to determine the physiological role of somatostatin in the pancreas and gut and all its actions on the alimentary system so far described suggest an overall peripheral regulatory function in the absorption of nutrients. For further reading on the central and peripheral roles of somatostatin see [5, 6, 10]. The action and therapeutic application of somatostatin analogues are considered in Chapter 12 Section 3(iii).

7:3 Neural Control of GH Secretion

(i) *Episodic GH Secretion*

The predominant pattern of GH secretion is intermittent and episodic, surges of GH release occurring about every 3-4 hours with a particularly prominent surge occurring during the first few hours of sleep, especially in young animals and children. This effect is triggered by sleep and is not simply a circadian rhythm effect characteristic of other hormones (see Chapter 9 Section 2). In humans the frequency and amplitude of this episodic release changes during puberty and the nocturnal elevation is reported to be absent in people older than 50. The levels of GH between surges is usually equivalent to or less than the lower limit of sensitivity of radioimmunoassays for GH (< lng/ml), while during the secretory bursts, GH levels in plasma may reach > 200 ng/ml in rat or > 50 ng/ml in man. The profile of secretion is acute, showing rapid rises and abrupt terminations, which is also consistent with the short half-life of the hormone. Amino acids such as arginine, leucine and isoleucine, and hypoglycaemia (low blood glucose) can stimulate GH secretion although fluctuations of glucose and other nutrients that occur normally in relation to food intake do not appear to alter or correlate with normal episodic GH secretion. Hyperglycaemia only temporarily suppresses daytime surges of GH release and nocturnal GH rises are not affected by fasting or hyperglycaemia although more frequent GH secretory pulses do occur during fasting. Thus, the relative inability of potential metabolic regulators to suppress GH secretion suggest that the episodic surges are the result of primary activation of GH secretion by hypothalamic neurosecretory cells.

The presence of two hypothalamic hormones regulating GH secretion has raised the questions as to whether the episodic secretion is simply due to a tonic inhibitory effect of somatostatin with intermittent rebound release or whether a releasing factor was a prerequisite for GH secretion. Based on the following evidence there is little doubt that GH secretion in the rat is regulated predominantly by a stimulatory hypothalamic influence. For details of localized lesions and stimulation experiments see Chapter 2 Section 3(ii). Lesions of the ventro-medial nucleus (VMN – see Figure 1.2) completely abolish episodic GH release and GH levels remain low whilst lesions of the preoptic anterior hypothalamic area results in almost complete loss of immunoreactive somatostatin in the median eminence with a rise in baseline levels of GH and an increased frequency of episodic GH release. Alternatively, stimulation of the VMN elicits prompt GH release while stimulation of the anterior hypothalamus and preoptic area inhibits GH secretion. Administration of a specific antiserum to somatostatin, in order to antagonise its activity, elevates basal GH release but does not interfere with the pulsatile pattern of secretion, and intermittent infusions of somatostatin cause inhibitory rebound of GH secretion only during episodic surges of GH secretion. From such data it has been concluded that the VMN is the source of GH-RF and that neurones in this region stimulate episodic GH secretion with a frequency between 1 and 3 hours. Superimposed on this system is the preoptic-

Figure 7.3: Schematic Diagram of the Proposed Dual Hypothalamic Control of Episodic Growth Hormone (GH) Secretion. The summated effects of growth hormone releasing factors (GHRF) and somatostatin (SOM) result in the episodic secretory pattern of GH. A particularly prominent GH surge occurs during the first few hours of sleep.

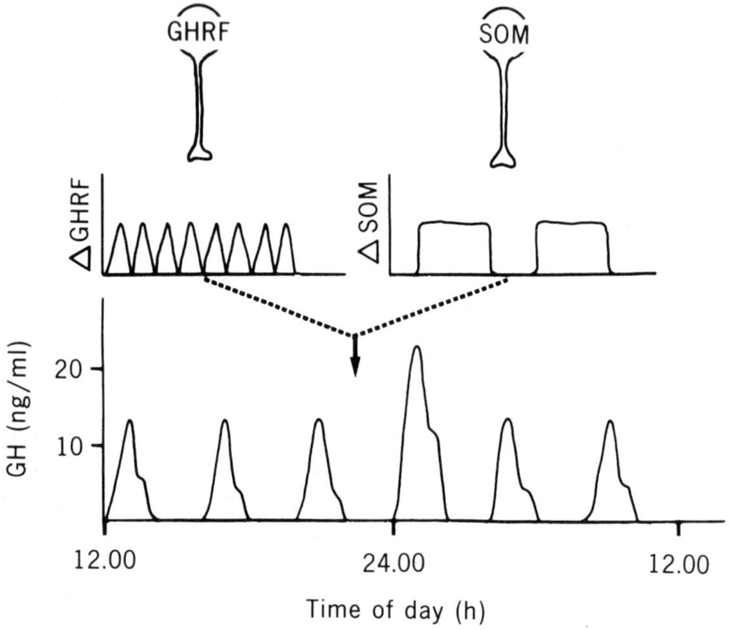

anterior hypothalamic secretion of somatostatin which causes only intermittent suppression of GH release possibly limiting any rebound effects and gives rise to the overall secretory pattern of GH hormone secretion as determined by peripheral blood sampling (see Figure 7.3).

(ii) *Physiological Factors Stimulating GH Secretion*

Exercise, many forms of stress (physical or emotional) as well as sleep (see Table 7.1) stimulate GH release in animals and humans, although certain forms of stress inhibit GH release in the rat. These responses suggest the involvement of extrahypothalamic areas in the control of this hormone. Stimulation of the hippocampus increases circulating levels of GH. In addition, stimulation of the basolateral amygdala elicits a prompt rise of GH secretion, an effect which is tightly entrained to the stimulus. On the other hand, stimulation of the corticomedial amygdala inhibits GH secretion, probably via a monosynaptic input which reaches both the VMN and the preoptic area via the stria

Table 7.1:

(A) Factors which increase GH secretion	
(1)	Stress: Physical and Psychological
(2)	Exercise
(3)	Sleep
(4)	Hypoglycaemia
(5)	Amino acids: arginine, leucine, isoleucine
(6)	Peptides: vasopressin, glucagon, substance P, neurotensin, opioid peptides
(7)	Monoamines, 5-hydroxytryptamine precursors, melatonin
(B) Factors which inhibit GH secretion	
(1)	Hyperglycaemia
(2)	Alpha-adrenergic blockers
(3)	Dopamine agonists (only in acromegaly)
(4)	Drugs blocking 5-hydroxytryptamine effects

terminalis. Complete deafferentation of the rat hypothalamus (see Chapter 2 Section 3(ii)), thereby depriving the VMN of any extra-hypothalamic input, enhances episodic secretion of GH. This observation, coupled with the fact that lesions of the preoptic area block the stress induced inhibition in GH in the rat suggests that extrahypothalamic inhibitory inputs to the basal hypothalamus are important to maintain tonic inhibition of GH secretion. For further details see [4, 5, 9].

(iii) *Feedback Regulation of GH Release*

Unlike the gonadotrophins, thyroid stimulating hormone and corticotrophin, which have specific target endocrine glands and secretion rates controlled by feedback loops, GH appears to act on most if not all cells in the body and there is no conclusive evidence for any feedback actions of peripheral hormones (factors?) secreted by target cells. Since the effects of GH on bone and cartilage are mediated by peptidergic factor(s), somatomedins, it has been suggested that somatomedins may be important as a feedback regulator of GH secretion (see Figure 7.4). There is, however, no good correlation between somatomedin activity in the plasma and the concentration of circulating GH, and it is not known how many other actions of GH, if any, may be mediated by somatomedin. Self regulation of GH by a short feedback loop has been proposed, as shown for prolactin (see Chapter 7 Section 5(iii)), but although there is evidence for GH inhibiting somatostatin secretion, proof for the significance of this effect is lacking.

Figure 7.4: Control of Growth Hormone Secretion: Possible Factors Involved. GH – Growth hormone; GHRF – growth hormone releasing factor; SOM – somatostatin; NA (α) – noradrenaline via α receptor; NA (β) – noradrenaline via β receptor; DA – dopamine; 5-HT – 5-hydroxytryptamine; POA/AHA – preoptic and anterior hypothalamic area; VMH – ventromedial nucleus; CM – cortico-medial amygdala; BL – basolateral amygdala.

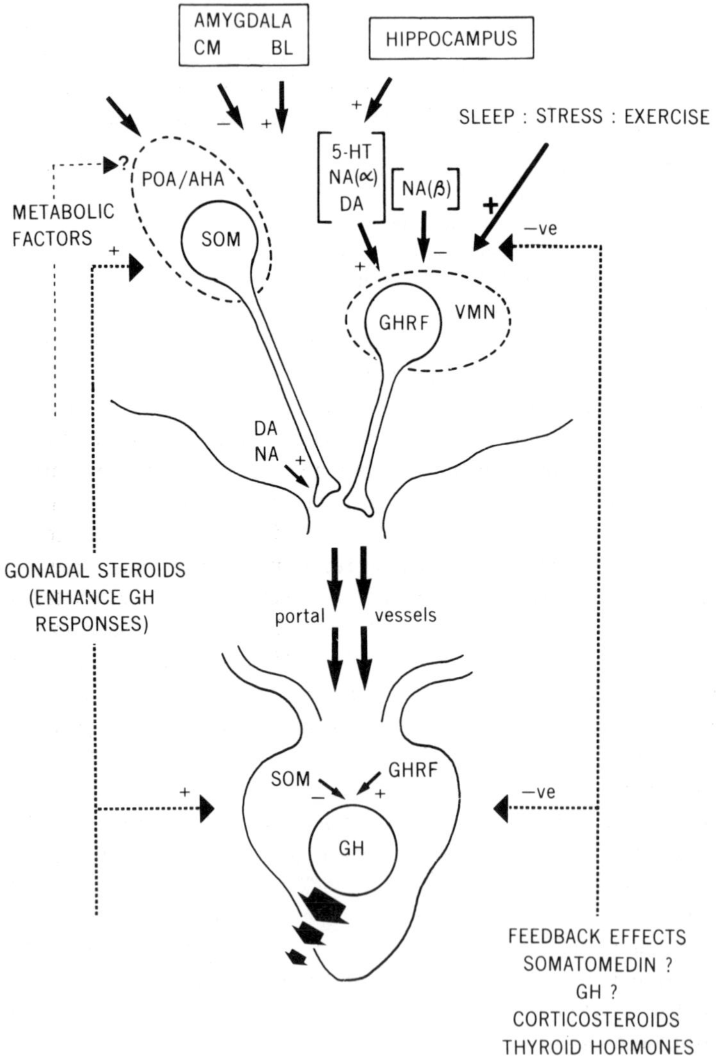

(iv) *Effects of Neurotransmitters, Neuropeptides and Hormones on GH Release*

The monoamines, dopamine, noradrenaline and 5-hydroxytryptamine (5HT) generally all enhance GH secretion (see Table 7.1). Pharamacological evidence suggests that GH release is facilitated by dopaminergic and noradrenergic α-receptor activation although a β-receptor mediated noradrenergic pathway may also be present (see Figure 7.4). The administration of the catecholamine precursor, L-dopa, which rapidly enters the brain to increase dopamine and noradrenaline levels, normally causes GH secretion, but paradoxically L-dopa and other dopamine agonist drugs have been used successfully for the treatment of acromegaly resulting in lowered levels of plasma GH concentrations (see Chapter 12 Section 1).

5-HT control may be related to the sleep-associated increase in episodic GH secretion since this neurotransmitter is an important mediator of sleep mechanisms (see Figure 7.4). Interestingly, *in vitro* studies have shown that dopamine and noradrenaline stimulate somatostatin release from median eminence whole tissue fragments but inhibit its release from synaptosomes whilst *in vivo* intraventricular injection of the monoamines increases somatostatin concentration in hypophyseal portal blood. These apparently confusing findings may suggest that generally the release of the GH-RF in response to these neurotransmitters may exceed that of somatostatin which is under a more subtle control, so that the net effect of these transmitters is to elevate circulating levels of GH. A wide variety of neuropeptides release GH including vasopressin, αMSH, substance P, neurotensin and the opioid peptides, enkephalins and endorphins (see Table 7.1) but whether they have any physiological role on the hypothalamic pituitary axis remains to be determined. Of interest is the potent stimulatory effect of TRH in causing GH release although this is only observed in acromegaly, and other pathological conditions including anorexia nervosa, renal failure and hypothyroidism (see Chapter 12 Section 1). Gonadal steroids, both oestrogens and androgens, enhance GH secretion and may account for sex differences in GH responses to arginine administration for example, whereas excess corticosteroids or thyroid hormones impair GH responses (see Figure 7.4). The pancreatic hormone, glucagon, also stimulates GH release. For further information see [5, 9].

Figure 7.5: Diagram of the Possible Factors which Interact at the Pituitary Lactotroph to Control Prolactin Secretion. PIF – prolactin inhibiting factor; PRF – prolactin releasing factor; DA – dopamine; GABA – γ-aminobutyric acid; TRH – thyrotrophin releasing hormone; VIP – vasoactive intestinal polypeptide; SOM – somatostatin; OP – opioid peptides.

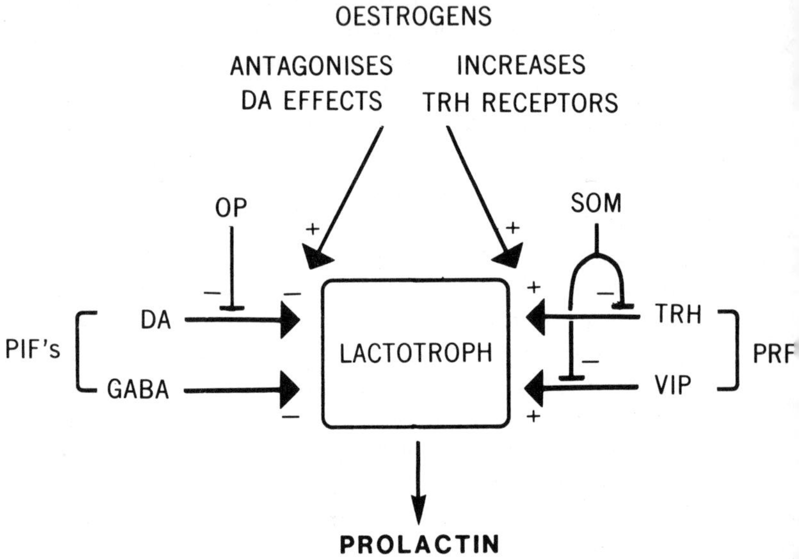

7:4 Hypothalamic Control of Prolactin Secretion – Prolactin Inhibiting and Releasing Factors (PIFs and PRFs)

In contrast with GH secretion, the release of prolactin from the adenohypophysis is under a tonic inhibitory control as illustrated by the fact that hypothalamic lesions elevate circulating levels of prolactin. However, prolactin secretion is influenced by various specific inhibiting and releasing factors even though in basal conditions the predominant hypothalamic influence is inhibitory. A major inhibitory factor that alters prolactin release is the neurotransmitter, dopamine, whereas the nature of a specific prolactin releasing factor remains to be defined. Control of prolactin secretion is very complex, however, and various other factors may be involved (see Figures 7.5 and 7.6). For example, thyrotrophin releasing hormone (TRH) and vasoactive intestinal polypeptide (VIP) are known to have powerful stimulatory

Figure 7.6: Possible Factors Involved in the Neural Control of Prolactin Secretion: Modulatory Feedback Effects of Hormones. PRL – prolactin; PIF – prolactin inhibiting factor; PRF – prolactin releasing factor; TRH – thyrotrophin releasing hormone; VIP – vasoactive intestinal polypeptide; DA – dopamine; NA – noradrenaline; 5-HT – 5 hydroxytryptamine; GABA – γ-aminobutyric acid; SCN – suprachiasmatic nucleus.

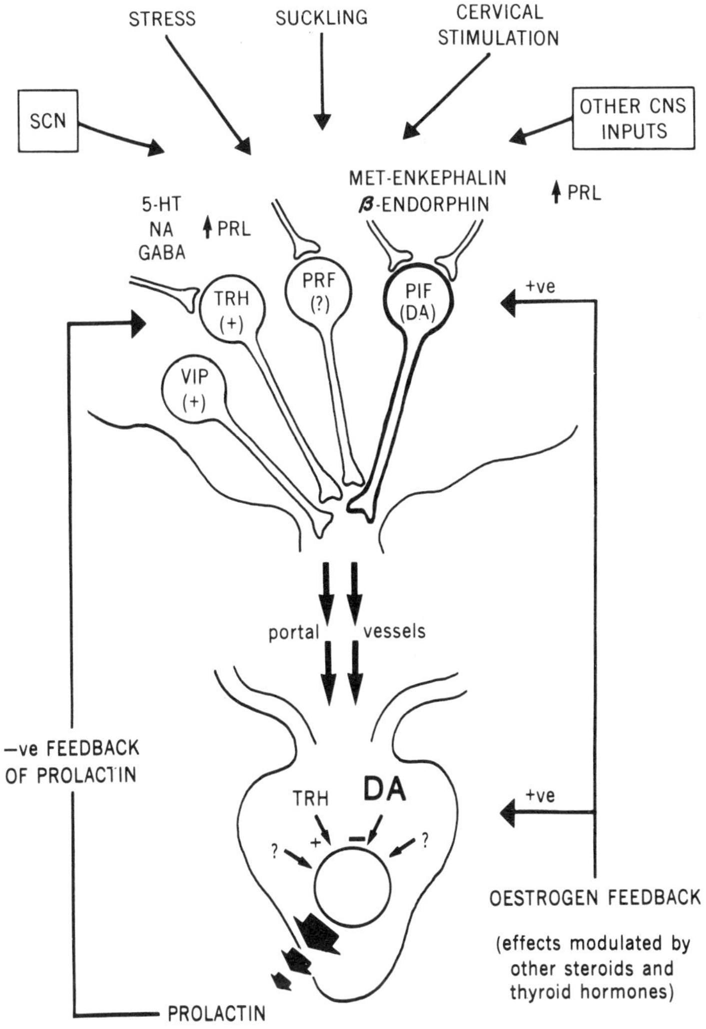

effects on prolactin release both *in vivo* and on isolated pituitary glands *in vitro* although the physiological importance of such effects has not been established.

(i) *Dopamine as the Prolactin Inhibiting Factor (PIF)*

The median eminence is rich in dopamine terminals of the tuberoinfundibular pathway. Some of these synapse with Gn-RH neurosecretory terminals in the lateral palisade zone and may directly regulate Gn-RH secretion (see Chapter 8 Section 6(i)). Other evidence, particularly from studies on median eminence tissue and isolated nerve terminals (synaptosomes), suggests that dopamine may also regulate TRH, corticotrophin releasing factor and somatostatin secretion in the median eminence. Other dopamine terminals in the median eminence, largely in the medial palisade zone, terminate on the hypophyseal portal capillaries and high levels (10^{-8}M) of dopamine have been measured in hypophyseal portal blood. Thus tuberoinfundibular dopamine may act at the pituitary level and studies with isolated anterior pituitary glands and cells demonstrate strong inhibitory actions of dopamine on the release of prolactin; these responses have been confirmed in intact animals. Pharmacological experiments with dopamine agonist and antagonist drugs led to the concept that this neurotransmitter may be the main and possibly the only prolactin inhibiting factor of consequence, and subsequent experimental data support and confirm this idea. For example, dopamine receptors are present in the anterior pituitary and are associated with isolated lactotrophs and other pars distalis cells. Also decreases in hypothalamic dopamine release occur in parallel with increases in prolactin secretion. In addition, uptake studies with radiolabelled dopamine in the median eminence have shown that the uptake of dopamine in the median eminence is lower than in the striatum, which is also rich in dopamine nerve terminals, suggestive of a neurosecretory function of hypothalamic dopamine nerve terminals. The importance of dopamine in the regulation of human prolactin secretion is exemplified by the widespread effective use of the dopamine agonist bromocryptine in the treatment of hyperprolactinaemia, galactorrhoea and infertility in women (see Chapter 12 Section 3).

Despite this evidence there remains some doubt as to whether dopamine is the only PIF, since hypothalamic extracts devoid of dopamine still show a dose-dependent inhibitory effect on prolactin secretion. In addition it has not been possible quantitatively to correlate the rate of dopamine secretion into the hypophyseal portal capillaries with the magnitude of the prolactin response. When prolactin secretion is stimulated it has been suggested that there are two phases to this response – initially a decrease in dopamine release (the preparative phase) followed by an increased secretion of a PRF. Such a suggestion raises the question as to whether a similar mechanism may exist in the inhibition of prolactin secretion i.e. a cooperative action of dopamine with another PIF.

Recently it has been shown that γ-aminobutyric acid (GABA) can inhibit prolactin secretion by an action on the anterior pituitary gland by an effect that does not involve dopamine receptors. However, the moderate maximal effects of GABA on the anterior pituitary as well as the limited amounts of this amino acid both in the median eminence and in the pituitary suggest that it does not play a major physiological role in the regulation of prolactin release (see Section 7:5). Until another PIF has been positively identified dopamine is considered to be the major inhibitory factor in controlling prolactin secretion. Its action on the pituitary gland is thought to be mediated by dopamine receptors of the D_2 (inhibitory) type which do not appear to be linked to an adenylate cyclase and changes in cyclic AMP but may inactivate Ca^{2+} channels thereby inhibiting release of prolactin.

(ii) *Prolactin Releasing Factors*

The physiological stimulation of prolactin release cannot be entirely accounted for by changes in the turnover rate of dopamine. For example, in lactating rats dopamine turnover is increased, despite the fact that prolactin secretion is raised, and recent evidence suggests a prolactin releasing factor (PRF) is present in the plasma of suckled women and lactating female rats. The nature of this PRF is unknown but both TRH and VIP potently stimulate prolactin secretion by an action on the anterior pituitary gland with an apparent affinity in the nanomolar range. Both peptides are present in significant amounts in hypophyseal portal blood and prolactin stimulation involves distinct recognition sites on the lactotrophs. The stimulatory effects of these peptides are additive and they can reduce the dopamine inhibition of prolactin release although it is unlikely that they act on the dopamine receptors. Recent studies have implicated yet further factors in prolactin control. Firstly, experiments indicate that somatostatin may inhibit the stimulatory effects of TRH and VIP, while further experiments have shown that Met-enkephalin and β-endorphin stimulate prolactin release although most evidence points to a hypothalamic site of action of opioid peptides (see Section 5). However, other studies have shown that such opioid peptides can antagonise the effects of dopamine on the pituitary *in vitro* by an independent receptor mechanism but are unable to stimulate the release of prolactin by themselves. Figures 7.5 and 7.6 summarise all the possible factors which may be involved in the control of prolactin secretion at the hypothalamic and pituitary levels. For further reading see [3, 8, 12].

7:5 Neural Control of Prolactin Secretion

(i) *Rhythms of Prolactin Release*

Like all other anterior pituitary hormones the secretion of prolactin is pulsatile in nature and plasma levels fluctuate between approximately 2 and 50 ng/ml plasma; its half life in plasma is approximately 15 minutes. In humans there is a sleep-induced increase in prolactin secretion, similar to that for GH, with a surge beginning about 1 hour after the onset of sleep reaching a peak about 5 a.m. in the morning. In female rodents diurnal surges in prolactin secretion may be induced by mating. In the rat cervical stimulation results in diurnal surges of prolactin secretion and lesion experiments have implicated the preoptic area (POA) and suprachiasmatic nucleus (SCN) to be involved in these twice daily surges of prolactin release. It has been proposed that the POA exerts a tonic inhibitory action on prolactin secretion which is removed by cervical stimulation; the SCN appears to control the timing of the episodic release in common with circadian rhythms (Chapter 9 Section 2). These diurnal surges terminate about the 10th day of pregnancy. Similar patterns of hormone secretion can be observed following sterile matings or mechanical stimulation of the cervix – this hormonal status is known as pseudopregnancy. Despite the fact that oestrogens are known to stimulate prolactin secretion, ovariectomy of a pseudopregnant rat does not abolish the diurnal secretory pattern but after day 5 or 6 only the nocturnal surge is present. The significance of the induced diurnal changes in prolactin secretion in rodents is not understood.

(ii) *Prolactin Release in Response to Suckling and Stress*

The important role of prolactin in lactation is well established and apart from the oestrogen stimulation of prolactin release neural inputs from the mammary gland are also able to induce prolactin secretion. During suckling, stimulation of sensory receptors in the nipple initiates a neuroendocrine reflex (similar to that described for oxytocin, Chapter

3 Section 8) and the intensity and duration of this stimulus is related to the quantity of prolactin secretion. This is associated with a known decreased release of dopamine and probably an increased secretion of a releasing factor (see Section 4(i)). Thus suckling is an important factor in the maintenance of lactation. Stress is another potent stimulus for prolactin secretion although the neural pathways eliciting this response have not been fully identified.

(iii) *Hormonal Modulation of Prolactin Secretion*

In general, target cells for prolactin are not endocrine glands which provide regulatory hormonal feedback signals. One exception is, of course, the gonads where prolactin plays a permissive role and can modulate the gonadotrophin receptor mechanism thereby altering the rate of steroid synthesis and release. It is well known that women have higher prolactin levels than men and this is probably due to the effects of oestrogens in stimulating both the synthesis and release of prolactin. Oestrogen appears to act at both the pituitary and the hypothalamic level. On the pituitary it stimulates synthesis of prolactin and has a powerful antidopaminergic effect which does not involve the dopamine receptor itself but presumably at some step following the binding of dopamine. Oestrogen treatment also increases the number of TRH receptor sites in the pituitary thereby potentiating the stimulatory action of TRH on prolactin secretion. The pituitary effects of oestrogen on prolactin release thus appear to involve both a decreased activity of the dopamine receptor and an increased sensitivity to TRH. On the hypothalamus, oestrogens can alter the activity of the tuberoinfundibular dopamine neurones and both studies on the turnover rate of hypothalamic dopamine and measurements of dopamine release into the hypophyseal portal capillaries show that raised levels of oestrogen decrease dopamine secretion. Thus oestrogens can raise prolactin release by an attenuation of the inhibitory hypothalamic dopamine control. However, it should be pointed out that under certain experimental conditions hypothalamic dopamine turnover is increased by oestrogen treatment but whether this is due to a direct effect of oestrogen on the hypothalamus or indirectly by raising prolactin levels which have an autoregulatory feedback effect on the dopaminergic neurones remains to be established. The effect of TRH on prolactin release (see Section 4) led to investigations on the possible role of thyroid hormone feedback, and while *in vivo* treatment with thyroxine does not appear to alter basal secretion of prolactin it can completely reverse the stimulatory effects of oestrogen. This inhibitory action may be mediated by changes in TRH receptor levels although an interaction between thyroid hormone and dopamine remains to be evaluated.

Apart from the feedback effects of oestrogens and thyroid hormones there is good evidence that prolactin may regulate its own secretion rate (autoregulation) (see Figure 7.6) by a short negative feedback loop (see Chapter 1 Section 9). When circulating levels of prolactin are raised, as occurs during lactation or by treatment with dopamine receptor antagonists (hyperprolactinaemia), the turnover rate of hypothalamic dopamine is increased as well as the concentration of this amine in the hypophyseal portal blood. This has led to the concept that circulating prolactin may control the release of dopamine from nerve terminals in the median eminence and this action may serve as a regulatory mechanism in controlling prolactin secretion. By the same action on hypothalamic dopamine turnover, prolactin may indirectly influence the secretion of other hypophysiotrophic hormones from the median eminence, e.g. Gn-RH and TRH. This may be particularly important in the control of Gn-RH secretion, since dopamine synapses on Gn-RH neurosecretory nerve terminals have been described and dopamine can inhibit Gn-RH secretion (see Chapter 8 Section 6(iii)). There is a high incidence of infertility associated with raised circulating levels of prolactin and it has been suggested that prolactin effects on hypothalamic dopamine may be the cause of amenorrhoea and anovulation observed in hyperprolactinaemia and reduced fertility during lactation. For further reading see [7, 8, 12].

(iv) *Neurotransmitters and Neuropeptides in the Hypothalamic Control of Prolactin Secretion*

Pharmacological studies have suggested a role for several neurotransmitters which may act on the hypothalamic neurosecretory cells controlling prolactin release. 5-hydroxytryptamine (5-HT) and noradrenaline have stimulatory effects on prolactin secretion, which is consistent with the effect of these transmitters on GH secretion. In the case of 5-HT there is evidence that it acts by stimulating the release of a PRF as opposed to inhibiting the release of dopamine. γ-aminobutyric acid (GABA) also exerts a stimulatory effect on prolactin secretion at the level of the hypothalamus (see Figure 7.6) and this effect appears more potent than its inhibitory action on the anterior pituitary. Agonists of GABAergic function are capable *per se* of lowering both tonic and oestrogen induced prolactin release, which suggests a functional role for GABA in stimulating prolactin secretion. It is unlikely that GABA effects are mediated by changes in hypothalamic monoamine function. The action of the endogenous opioid peptides, Met-enkephalin and β-endorphin, on prolactin secretion are well established. Their central administration causes an acute inhibition of dopamine turnover resulting in an increased release of prolactin. Furthermore, the stress or suckling induced prolactin surges are abolished by treatment with naloxone (an opioid receptor antagonist) suggesting that endogenous opioids are involved in these responses. Whether the main opioid responsible for this mediation is β-endorphin, which is probably secreted from the anterior pituitary gland along with ACTH during stress (see Chapter 4 Section 5) or neuronal Met-enkephalin released from the median eminence, is undetermined at the present time.

References

1. Dannies, P. S. (1980) Prolactin. *Trends in Pharmacological Sciences*, **1**: 206-8.
2. Frantz, A. G. (1978) Prolactin. *New England Journal of Medicine*, **298**: 201-7.
3. Macleod, R. M. (1976) Regulation of Prolactin Secretion, in L. Martini & W. F. Ganong (eds) *Frontiers in Neuroendocrinology* (Raven Press, New York), vol. 4, pp. 169-194.
4. Martin, J. B. (1976) Brain Regulation of Growth Hormone Secretion, in L. Martini & W. F. Ganong (eds) *Frontiers in Neuroendocrinology*, (Raven Press, New York), vol. 4, pp. 124-68.
5. Martin, J. B., Brazeau, P., Tannenbaum, G. S., Willoughby, J. O., Epelbaum, J., Terry, L. C. & Durand, D. (1978) Neuroendocrine Organization of Growth Hormone Regulation, in S. Reichlin, R. J. Baldessarini & J. B. Martin (eds) *The Hypothalamus* (Raven Press, New York), pp. 329-71.
6. McCann, S. M., Krulich, L., Negro-Vilar, A., Ojeda, S. R. & Vijayan, E. (1980) Regulation and Function of Panhibin (Somatostatin) in E. Costa & M. Trabucchi (eds) *Neural Peptides and Neuronal Communication* (Raven Press, New York), pp. 131-43.
7. McNeilly, A. (1979) Effects of Lactation on Fertility. *British Medical Bulletin*, **35**: 151-4.
8. Nicoll, C. S. (1980) Prolactin (symposium). Series of articles in *Federation Proceedings*, **39**: 2561-98.
9. Pecile, A. & Olgiati, V. R. (1978) Control of Growth Hormone Secretion, in S. L. Jeffcoate & J. S. M. Hutchinson (eds) *The Endocrine Hypothalamus* (Academic Press, New York), pp. 362-83.
10. Reichlin, S. (1980) Peptides in Neuroendocrine Regulation in F. E. Bloom (ed.) *Peptides: Integrators of Cell and Tissue Function* (Raven Press, New York), pp. 235-50.
11. Tanner, J. M. (1981) *Control of Growth* (Churchill Livingstone, Edinburgh).
12. Tindal, J. S. (1978) Control of Prolactin Secretion, in S. L. Jeffcoate & J. S. M. Hutchinson (eds) *The Endocrine Hypothalamus* (Academic Press, New York), pp. 333-60.
13. Van Wyk, J. J. & Underwood, L. E. (1980) Growth Hormone, Somatomedins and Growth Failure, in D. T. Krieger & J. C. Hughes (eds) *Neuroendo-*

crinology – a Hospital Practice Book (Sinauer Assoc. Inc., Sunderland, MA), pp. 229-309.

14. Zingg, H. H. & Patel, Y. C. (1979) Somatostatin Precursors: Evidence for Presence in and Release from Rat Median Eminence and Neurohypophysis. *Biochemical and Biophysical Research Communications*, 90: 466-72.

8

CONTROL OF GONADOTROPHIN HORMONE SECRETION

8:1 The Hypothalamic-Pituitary-Gonadal Axis **8:2** The Gonadotrophins (i) Structure (ii) Measurement of LH and FSH (iii) Action of LH and FSH on the gonads **8:3** Gonadotrophin Releasing Hormone (Gn-RH) (i) Structure, synthesis and degradation and synthetic analogues (ii) Localisation of Gn-RH containing neurones (iii) Action on the pituitary gonadotrophs **8:4** Feedback Effects of the Gonadal Steroids on Gn-RH and Gonadotrophin Secretion **8:5** Sex-Steroid Receptors in the Brain **8:6** Neural Regulation of Gonadotrophin Secretion (i) Neural inputs to the Gn-RH neurosecretory cells (ii) Electrophysiological correlates of gonadotrophin secretion (iii) Putative neurotransmitters (iv) Neuropeptides and gonadotrophin secretion

8:1 The Hypothalamic-Pituitary-Gonadal Axis

The gonads (ovaries or testes) are the target glands for the gonadotrophins, luteinising hormone (LH) and follicle stimulating hormone (FSH), and the processes of ovulation, spermatogenesis and the secretion of sex steroids are all regulated by the action of these two anterior pituitary hormones on the testes and ovaries. The sex steroids, the male androgens (e.g. testosterone), the female oestrogens (e.g. oestradiol) and progesterone (see Figure 5.1) have diverse actions, not only on the reproductive systems in maintaining normal function, but also on sexual differentiation, growth and development. In the foetus differentiation of the reproductive tracts and the brain (see Chapter 11 Section 3(ii)) are controlled by steroids secreted by the developing gonads, and subsequent sexual development is dependent on gonadal steroid secretions. The feedback actions of the sex steroids on the brain and pituitary are important in the regulation of gonadotrophin secretion and sexual behaviour in the adult and in determining the onset of puberty. Sex steroids are secreted in relatively small amounts from the adrenal cortex and, although the significance of this particular source of sex steroid secretion under normal circumstances is not fully understood, hypersecretion in the adrenal cortex can lead to physiological changes associated with abnormally high levels of circulating sex steroids. Many excellent reviews have been written on the actions of sex steroids and the control of gonadal function and the reader is referred to [2, 4, 8].

The rate of secretion of the gonadotrophin hormones is controlled by gonadotrophin releasing hormone (Gn-RH), a peptide whose structure is known. Gn-RH is released from hypothalamic nerve terminals into the hypophyseal portal vessels and transported to the anterior pituitary where it stimulates the synthesis and release of LH and FSH. There has been much debate as to whether there are two releasing factors, one specific for LH and the other for FSH but, to date, only one has been identified and it is widely thought that the differential release of LH and FSH may be determined by other factors. The major hormones involved in the hypothalamic pituitary gonadal axis are summarised in Figure 8.1 together with the possible factors which control gonadal function. This chapter will be primarily concerned with the central mechanisms which regulate Gn-RH, LH and FSH secretion and the complex integration of neural inputs and feedback effects of sex steroids which determine the central control of ovulation and spermatogenesis. Reference will also be made to prolactin in this chapter. Although this pituitary hormone is best known for its role in lactation (see Chapter 7 Section 1(iv)) it should be considered as a gonadotrophin because it has been shown to play a role in the regulation of luteal function in rodents, sheep, and probably humans.

The changes in the reproductive system which occur at the time of puberty have not been considered in this chapter but it is alluded to in Chapter 9 Section 3(iii) since it has been suggested that the process of puberty shows many analogies to central mechanisms controlling seasonal breeding. Reference [12] gives further details of the dramatic changes which occur at the time of sexual maturation.

8:2 The Gonadotrophins

(i) *Structure*

LH and FSH are two of a distinct family of glycoprotein hormones secreted from the anterior pituitary gland; thyrotrophin (TSH) being the other adenohypophyseal glycoprotein. A fourth glycoprotein is chorionic gonadotrophin, a placental hormone found only in pregnant females of higher mammals. The molecular weights of the gonadotrophins are in the 30,000–35,000 range. The hormones are secreted by basophil cells of the adenohypophysis and although histological

Figure 8.1: The Hypothalamic-Pituitary-Gonadal Axis. Gn-RH – gonadotrophin releasing hormone; LH – luteinising hormone; FSH – follicle stimulating hormone; ARC – arcuate nucleus; AC – anterior commissure; DM – dorsomedial nucleus; MB – mammillary bodies; OC – optic chiasma; POA – preoptic area; PV – paraventricular nucleus; SC – suprachiasmatic nucleus; VM – ventromedial nucleus; + – positive feedback effects; – – negative feedback effects.

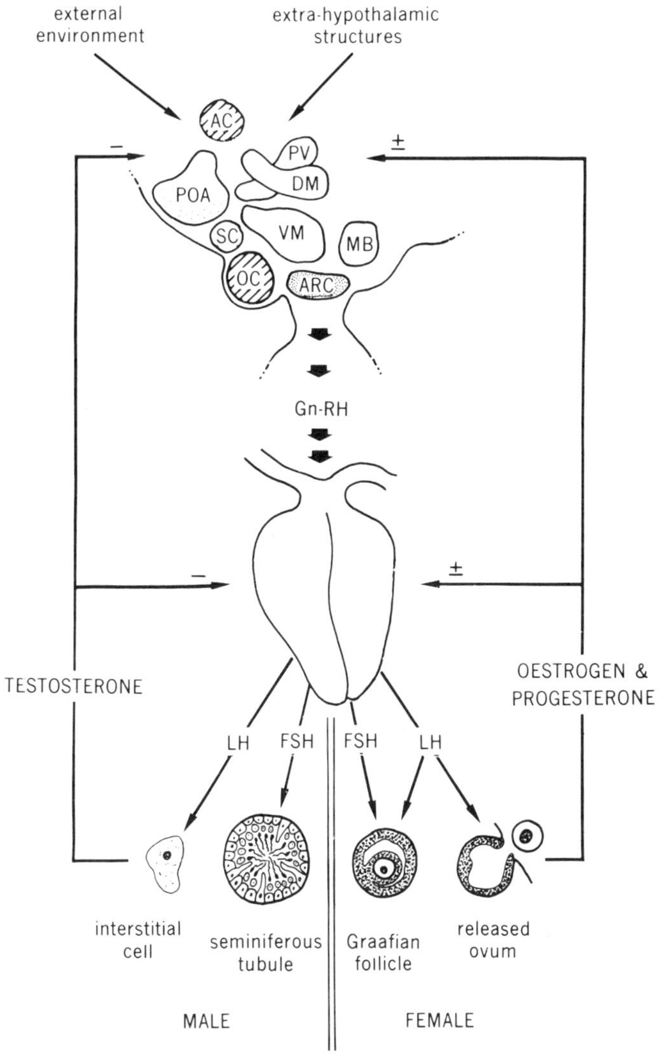

distinction can be made between thyrotrophs and gonadotrophs, there is still some dispute as to whether one can define separate LH and FSH containing basophil cells. These hormones share many functional and structural properties, each hormone being made up of two non-identical polypeptide chains, the α- and β-subunits, and a carbohydrate moiety which represents between 6 and 30 per cent of the total molecular mass. The amino acid sequences for both the α- and β-subunits of the glycoproteins have been determined in several species including pig, sheep, cow and human and the α-units of the four glycoprotein hormones show remarkable structural homology within species whereas the β-units differ markedly and confer biological specificity to the hormones: both subunits are required for hormone activity. Dissociation of the α and β subunits can be achieved by various agents and the formation of hybrid molecules containing the α unit of one hormone and the β unit of another leads to reconstitution of the activity of the hormone from which the β unit was derived. For example, LH β-subunit + TSH α-subunit gives rise to a hormone with LH activity. Similar recombination experiments have been carried out for FSH, TSH and chorionic gonadotrophin. (For further information on glycoprotein hormone structure and function see [7] in Chapter 6.)

(ii) *Measurement of LH and FSH*

One of the earliest bioassays for LH was the assessment of ovulation by examining the Fallopian tubes for the presence of released ova. Although this did not provide a quantitative estimation of LH secretion it did indicate the occurrence of the preovulatory surge of LH and was a particularly easy and useful index of anterior pituitary secretion, often used in early neuroendocrine studies prior to the development of radioimmunoassay techniques. *In vitro* bioassays for LH and FSH are similar to those described for ACTH in that they involve measuring gonadotrophin-induced steroid synthesis or secretion in isolated gonadal tissue or measurement of ovarian ascorbic acid depletion to assess LH activity. Steroid secretion is quantified by radioimmunoassay procedures. Common bioassays currently in use are the LH stimulation of testosterone secretion from isolated Leydig cells of rat testes and the FSH-stimulated conversion of androgens into oestrogens in isolated Sertoli cells of the testes. These bioassays are generally more sensitive than the respective radioimmunoassays and do measure biological activity as opposed to immunological activity. However, the procedures are lengthy and not very practicable when large numbers of samples are involved. Radioimmunoassays for both LH and FSH are sufficiently sensitive to measure circulating levels of the hormones whose concentration is in the order of 50 pg/ml plasma. The specificity of the antisera resides in its ability to detect differences in the amino acid sequence of the respective β-subunits, so that cross-reactivity with the other gonadotrophin or TSH is minimal. This is not true for LH and chorionic gonadotrophin (CG) since antisera raised to one hormone cannot distinguish between the two hormones, which are structurally very similar and have identical effects on the gonads where they appear to compete for the same receptors.

(iii) *Action of LH and FSH on the gonads*

The receptors for the gonadotrophins (as for all glycoprotein hormones) are proteins located on the plasma membrane of target cells. They have a high affinity for their respective hormones, with dissociation constants in the order of 10^{-10}M. Both FSH and LH act on their gonadal target cells by stimulating adenylate cyclase which causes an increase in the intracellular concentration of cyclic AMP (see Chapter 1 Section 10). Prostaglandins have also been implicated in regulating certain gonadal functions, perhaps most notably luteolysis (breakdown of the corpus luteum). However, it is now considered that the effects of prostaglandins may act independently of LH-receptor mechanisms.

The primary target for LH in the testes are the Leydig cells where it stimulates testosterone production. The latter hormone is essential for spermatogenesis and the maintenance of Sertoli cell function (see below). In the ovary, LH binds with all the tissue components of the gland – namely the interstitial cells, follicular cells (theca interna cells and granulosa cells) and corpora luteal cells – and in all cases the effects of LH are to increase steroidogenesis of oestrogens (androgens) and progesterone. The relative proportions of these steroids which are synthesised will depend upon the stage of development of the follicle.

Binding studies of FSH in the gonads have indicated that FSH receptors are located on only one cell type in either the testis or ovary – namely the Sertoli cells or the granulosa cells respectively. In the testis the Sertoli cells play an important role in spermatogenesis providing nutrients for the developing sperm and forming a unique blood–testis barrier which serves to protect the maturing sperm. The action of FSH on these cells is to stimulate the production of an androgen binding protein (the physiological significance of this binding protein is unclear), to stimulate the conversion of testosterone into oestrogen, and to initiate fluid movement into the seminiferous tubules. In the granulosa cells FSH stimulates the conversion of androgens into oestradiol and passage of follicular fluid into the antrum where the ovum is developing. In addition it also stimulates protein synthesis and mitotic activity of the granulosa cells.

In the female LH is also responsible for resumption of meiosis of the oocyte and for rupture of the follicle causing ovulation. However, one of the primary physiological effects of the gonadotrophins is stimulation of the steroidogenesis, and this is likely to be mediated by stimulation of enzymes which either increase the amount of free cholesterol available for steroid synthesis or stimulate the conversion of cholesterol into pregnenolone or activate enzymes at other points in the steroidogenic pathway (see Figure 5.1).

It has been known for some time that many hormones and drugs can regulate their own receptors by a mechanism analogous to that for receptor binding drugs and that chronic exposure of target cells to high concentrations of a hormone or drug can cause a negative or 'down' regulation of the production of the receptors (see Chapter 1 Section 10). Recently it has been demonstrated that LH may also regulate its own receptors since exposure of animals to high concentrations of LH *in vivo* causes, within hours, a parallel loss of LH binding and stimulation of adenylate cyclase. The significance of such an effect in the male is not clear although in the female it may play an important role in luteolysis (breakdown of the corpus luteum). LH receptors may also be modulated by other hormones and both FSH and prolactin may regulate LH receptor number and/or affinity. For further reading see [3, 7, 8, 9, 14, 15, 16].

8:3 Gonadotrophin Releasing Hormone (Gn-RH)

(i) *Structure, Synthesis and Degradation and Synthetic Analogues*

Gn-RH (LH-RH) was the second hypothalamic hypophysiotrophic hormone to be isolated and characterised and its structure was first published in 1971. Schally and his collaborators identified the decapeptide by working on 800 μg of purified hormone isolated from 160,000 pig hypothalami. A few months later Guillemin's group announced an identical structure for sheep Gn-RH: pyro Glu-His-Trp-Ser-Tyr-Gly-Leu-Arg-Pro-Gly-NH_2 (see also Figure 12.3). Identification of the amino acid structure of this hypothalamic hormone led to the production of synthetic Gn-RH and it thus became available for clinical and research purposes. Synthetic Gn-RH also enabled the raising of antibodies to the hormone so that both radioimmunoassays and immunohistochemical techniques for measuring and detecting the peptide could be established. There have been some reports of measurements of Gn-RH hormone in the peripheral blood circulation but generally the circulating concentration is too low to be accurately assessed. Therefore direct measurements of the secretory pattern of Gn-RH requires radioimmunoassay of hypophyseal portal blood which is collected by cannulating the pituitary stalk.

Little is known of the biosynthesis of this hormone although there is some evidence that Gn-RH is synthesised in a manner similar to other peptide hormones i.e. a larger prohormone is first synthesised on the ribosomes and then broken down by enzymes to the biologically active hormone.[11] Gn-RH is degraded and inactivated by both tissue and plasma peptidase enzymes which cleave the peptide at various sites. These sites of cleavage are associated with enzymes which recognise specific amino acid linkages and act on all peptides with such linkages. For example the peptidase which cleaves the pyro Glu-His of the Gn-RH molecule can also act on this same linkage in TRH (see Chapter 6 Section 3(v)). Studies on the degradation of Gn-RH have mainly been directed toward the mechanisms which inactivate the peptide at its site of synthesis in the hypothalamus and its site of action at the pituitary. It has been reported that an important inactivating mechanism is an internal cleavage of the Gn-RH molecule by an endopeptidase forming Gn-$RH_{1\text{-}6}$ and Gn-$RH_{7\text{-}10}$, although more recent evidence suggest that sites Tyr_5-Gly_6 and Trp_3-Ser_4 may also be important. Cleavage of the peptide at many more sites has been demonstrated, and some of the peptidase enzymes identified, but at the present time it is not possible to relate these biochemical findings directly to physiological mechanisms for the inactivation of Gn-RH. There is some evidence that tissue peptidases may regulate the amount of Gn-RH in the hypothalamus which is available for release. It has also been suggested that one mechanism of steroid feedback may involve modulation of peptidase activity in the hypothalamus but, in general, experiments have failed to substantiate this idea fully.

Peptidase enzymes are also present in plasma and the major sites of cleavage of Gn-RH added to plasma have been reported to be at the Trp_3-Ser_4 and Tyr_5-Gly_6 linkages. The significance of Gn-RH degradation by blood peptidases is unknown since it appears that the kidney is the major site of degradation *in vivo*. The circulating half life of Gn-RH is in the order of 5-8 minutes whereas the *in vitro* half life of Gn-RH in plasma is much longer and is in the order of hours. Such findings argue against the importance of blood peptidases although the extrapituitary action of Gn-RH may place greater significance on plasma degradation.

Studies using fragments of the Gn-RH molecule have shown that the complete molecule is required for the biological activity on LH and FSH secretion since removal of any one of the amino acids results in negligible release of LH and FSH. Conversely substitution of one or more amino acids produces analogues of Gn-RH with either agonistic or antagonistic properties, and to date several hundred analogues have been synthesised with a view to producing therapeutic drugs for either the stimulation or inhibition of fertility (see Chapter 12 Section 3(ii)). The required properties for antagonist analogues of Gn-RH are high affinity for the Gn-RH receptors, prolonged occupation of the receptors (thereby excluding endogenous Gn-RH) and as little, or no, biological activity. The best antagonists have resulted from substitutions in the His^2, Trp^3 and Gly^6 positions, (see Figure 12.3) although there is still a requirement for more potent antagonists to be developed. Interestingly, chronic administration of agonist analogues can produce antifertility effects and this finding has been turned to advantage by investigating the possible use of Gn-RH agonists as potential contraceptives.

(ii) *Localisation of Gn-RH Containing Neurones*

Immunohistochemical and radioimmunoassay studies have demonstrated the existence of two groups of neurones containing Gn-RH in several mammalian species: one in the medial basal hypothalamus (the arcuate nucleus) which projects to the external layer of the median eminence and a second group in the medial preoptic and medial septal nuclei which project primarily to the organum vasculosum of the lamina terminals (OVLT). However, more recent immunohistochemical studies have now shown that most, if not all, Gn-RH containing nerve cell bodies lie in the more rostral regions of the hypothalamus and preoptic area and not in the basal hypothalamus as was originally proposed. These divergent results have been attributed to the lack of specificity of the Gn-RH antisera used in previous investigations. It is interesting that the distribution of Gn-RH in the central nervous system, and indeed in other body tissues, appears to be more restricted compared with the other hypophysiotrophic peptides, TRH and somatostatin, using similar immunological procedures (see Chapter 10 Section 2). Very recently a Gn-RH-like peptide has been located in the ovary and the finding that Gn-RH and some analogues inhibit ovarian steroidogenesis and follicular development has prompted speculation on a possible peripheral role of Gn-RH outside the hypothalamic-pituitary axis.

(iii) *Action on the Pituitary Gonadotrophs*

It is generally agreed that the release of Gn-RH from nerve terminals is by a process of neurosecretion (see Chapter 3 Section 6) and that the immediate cellular events following peptide–receptor binding on the adenohypophyseal gonadotrophs is an increase in intracellular Ca^{2+} and cyclic AMP as described for the action of LH itself (see Chapter 1 Section 10). The biological responses induced by the action of Gn-RH on the pituitary are an increased release and synthesis of both LH and FSH. There has been considerable debate as to whether or not Gn-RH (also termed luteinising hormone releasing hormone, LH-RH) is specific for LH and that a separate releasing factor exists for FSH. Generally the evidence is mostly against this idea, despite the fact that under certain physiological conditions, e.g. puberty and during the menstrual cycle, there is a change in the circulating LH:FSH ratio. It has been proposed that differential effects of Gn-RH on LH and FSH release are due to feedback mechanisms of steroids which can alter both the rate of Gn-RH secretion and the response of the pituitary gonadotrophs. A further factor which could also regulate the LH:FSH ratio is 'inhibin', a peptidergic substance obtained from ovarian and testicular extracts. Inhibin causes a preferential decrease in FSH secretion although it still remains to be characterised and its physiological function remains in doubt. For further information see [5].

Gn-RH is secreted in a pulsatile fashion, as are the gonadotrophins, and the rate and amplitude of such pulses appear to be of paramount importance in determining the magnitude of the response of the pituitary. Since Gn-RH stimulates gonadotrophin synthesis it is likely that the amount of Gn-RH reaching the pituitary will regulate the content of LH and FSH and therefore determine the amount of gonadotrophins that are available for release. From studies in which LH responses to a continuous infusion of Gn-RH have been measured it has been proposed that there are two pools of LH in the pituitary – a readily releasable pool and a reserve pool. The relative size of these pools will determine both the amplitude and duration of the response to a single injection of Gn-RH (see Figure 9.3), and the LH responses do indeed vary throughout the menstrual or oestrous cycle as a consequence of changes in Gn-RH secretion and a direct effect of steroid feedback at the level of the pituitary (and Figure 8.2a). This releasing hormone also has a 'self-priming' action on the gland, in that repeated

Figure 8.2: (a) LH Responses to Increasing Doses of Gn-RH at Different Stages of the Rat Oestrous Cycle. Prior to ovulation (proestrus) oestrogen levels are high, the pituitary is more responsive to Gn-RH and the amplitude of the responses indicates an increase in the releasable pool of LH in the pituitary. On dioestrus day 1 oestrogen levels are low and LH responses are attenuated. (b) Self-priming effect of Gn-RH. Repeated stimulation of the anterior pituitary gland with sequential pulses of Gn-RH (100 ng/ml) increases the LH responses.

(a)

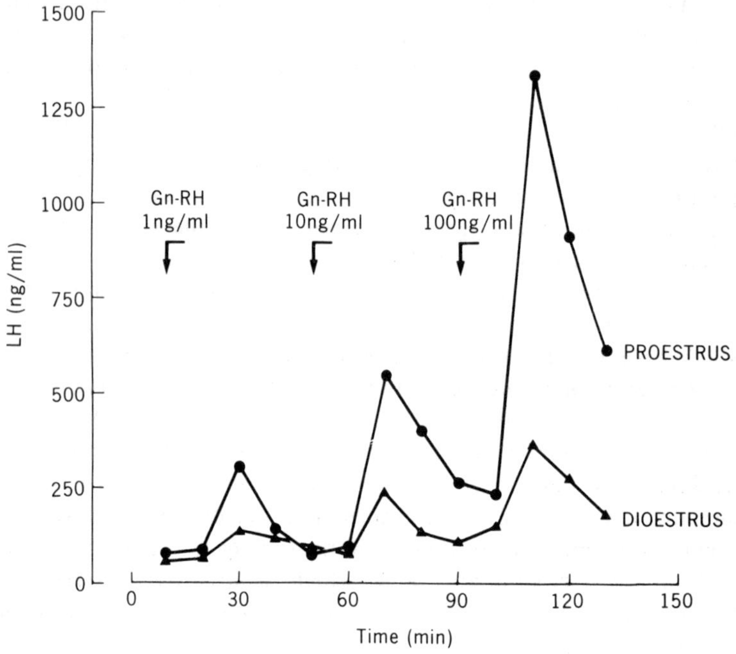

(b)

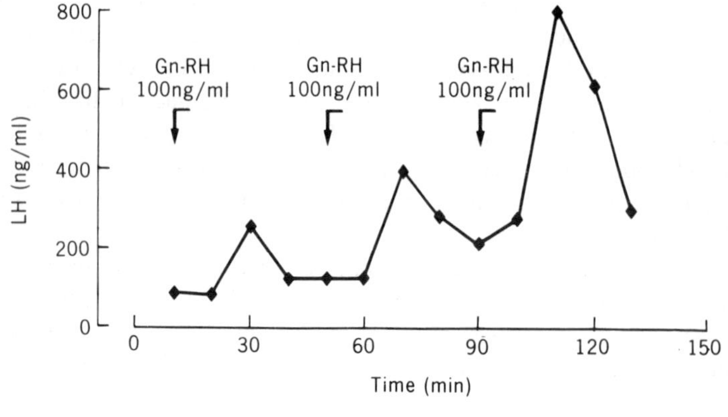

bolus injections of Gn-RH results in sequential increases in the amplitude of the LH responses (see Figure 8.2b) although a similar action on FSH responses has not been described. Furthermore, several factors may regulate Gn-RH receptors in the pituitary. There is evidence for steroid modulation of receptor density since the Gn-RH binding capacity of the pituitary has been shown to change during sexual development, during the oestrous cycle in rats and after removal of the gonads. In addition self regulation of Gn-RH receptors has been implicated since hyperstimulation by the native peptide or a long-acting analogue results in decreased responsiveness and such 'down regulation' is associated with a loss or internalisation of receptors (see Chapter 1 Section 10). Now it is generally accepted that the rate of endogenous Gn-RH release, in conjunction with steroid feedback mechanisms, controls the pituitary Gn-RH receptors. For further reading on Gn-RH see[8, 10, 11, 14]. The actions and therapeutic applications of Gn-RH analogues are considered in Chapter 12 Section 3(ii).

8:4 Feedback Effects of Gonadal Steroids in the Control of Gonadotrophin Secretion

One of the major mechanisms controlling the secretion of gonadotrophins is the feedback effects of sex steroids both at the brain and the pituitary gland (Figure 8.3). In the adult male the rate of gonadotrophin secretion does not show cyclical changes and remains relatively constant due to the negative feedback effects of testosterone. However, the mature female shows a cyclical rhythm of gonadotrophin release and oestrogens have been shown to exert both positive and negative feedback effects. The positive feedback action occurs mid-cycle when oestrogen levels are high due to increased secretion from the maturing follicle(s). These relatively high levels of oestrogen stimulate what is known as the preovulatory LH surge (positive feedback) and this surge is responsible for rupture of the mature follicle and release of the ovum (ovulation). The hormonal events which occur during the oestrous or menstrual cycle are diagrammatically represented in Figure 8.4 and the reader is referred to [8, 13, 14] and also [2] in Chapter 1 for further details.

Generally at the pituitary level high concentrations of oestrogens are known to increase the amount of LH secreted in response to Gn-RH (as occurs prior to ovulation) whereas testosterone inhibits Gn-RH effects on pituitary LH release. Progesterone appears to have a biphasic

Figure 8.3: Feedback Effects of Sex Steroids on the Hypothalamus and Anterior Pituitary Gland. DA – dopamine; NA – noradrenaline; 5-HT – 5-hydroxytryptamine.

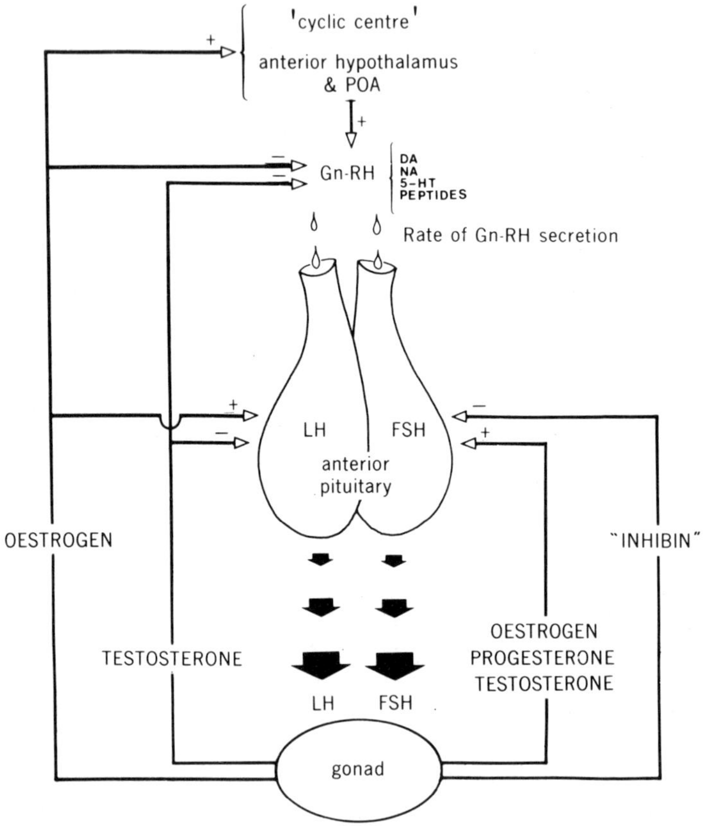

effect, initially promoting Gn-RH responsiveness followed by an inhibitory effect. In contrast all three steroids have an exclusively stimulating action on the release of FSH in response to Gn-RH (Figure 8.3). Apart from steroid modulation of the Gn-RH response in the anterior pituitary, there is also substantial evidence showing that gonadal steroids can act at the level of the hypothalamus to alter the rate of Gn-RH secretion in both sexes. In the female the preovulatory surge of LH is associated with an increased concentration of Gn-RH in hypophyseal portal blood (thought to result from an increased frequency and/or amplitude of the pulsatile Gn-RH secretion) and in the rat gonadectomy results in a rise of both Gn-RH and gonado-

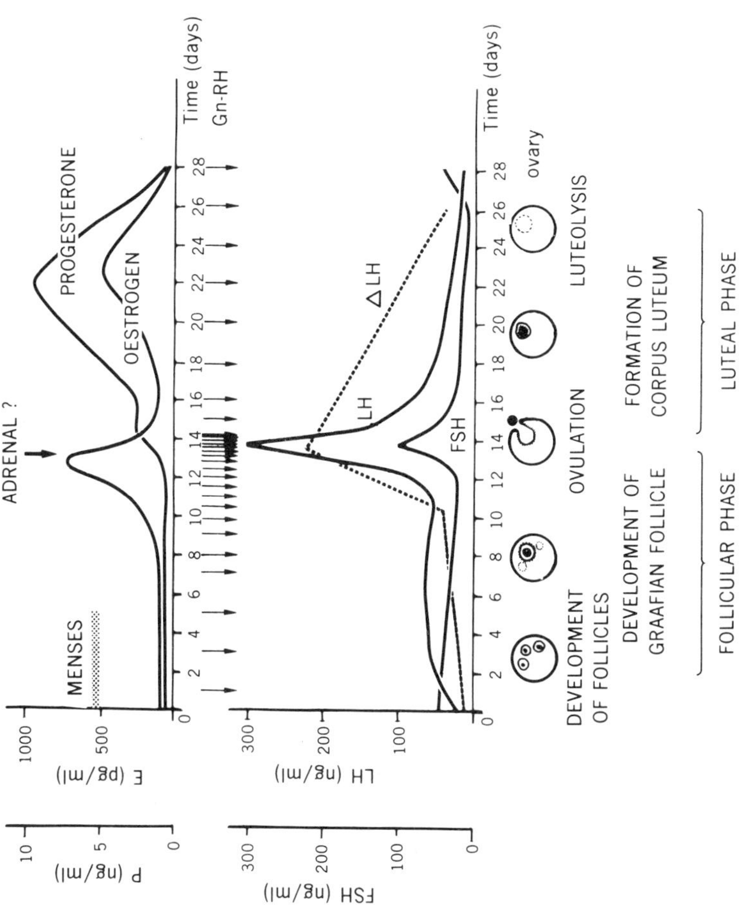

Figure 8.4: Hormonal and Ovarian Changes throughout the Menstrual Cycle. Arrows indicate the postulated changes in the frequency of episodic gonadotrophin releasing hormone (Gn-RH). Adapted from Fink, G., (1979) Brit. Med. Bull. 35:155, with permission.

trophin secretion due to the loss of negative feedback of steroids.

Clearly the mechanisms which trigger the preovulatory surge of LH are complex and appear to involve steroid modulation of pituitary responses to Gn-RH and stimulation of Gn-RH release by an action on the hypothalamus. Since an increase in the rate of Gn-RH release can itself modulate pituitary responses, distinguishing between different components of the steroid hormone feedback effects are particularly difficult. From studies on steroid feedback effects and measurement of Gn-RH secretion it has been proposed that the increase in pituitary responses observed prior to and during the LH surge occurs in two phases. The first occurs gradually due to the elevated levels of oestrogen and the second phase occurs rapidly and abruptly (10-fold) just prior to ovulation, perhaps largely due to the self-priming capacity of Gn-RH. Progesterone can also augment the effects of oestradiol but, because of the timing of the progesterone surge in many species, it is not certain whether this plays a major role in stimulating ovulation. Clearly we have great problems in determining whether the hypothalamus or the pituitary is the major site of steroid feedback, and although there is obviously a neural component of feedback action in several mammalian species, studies on monkeys have suggested that the pituitary is the major site. Nevertheless, in primates and rodents an LH surge can be induced in ovariectomised animals by oestradiol implants and it appears that, provided the hypothalamic-pituitary axis is exposed to sufficient levels of oestrogen for a certain time period (hours to days depending on species and length of cycle), gonadotrophin secretion is stimulated by a central positive feedback action. However, it should be mentioned that when high concentrations of oestrogens in the presence of progesterone are maintained, LH and FSH secretion are inhibited and this effect is the basis of the oral contraceptive pill which inhibits the occurrence of ovulation.

8:5 Sex-Steroid Receptors in the Brain

Steroid target tissues are widely distributed throughout the body and include the gonads and reproductive tracts, the accessory sex organs, liver, muscle and the central nervous system. The interaction of steroids with their receptors induces a wide spectrum of tissue-specific responses ranging from growth and differentiation, enzyme induction, anabolic responses and of course maintenance of spermatogenesis and ovulatory cycles. It is beyond the scope of this book to give a full account of all steroid receptor interactions and only those concerned with the central feedback action of gonadal steroids are described (see Chapter 1 Section 11). However, it should be pointed out that much of our current knowledge of steroid receptors is based on studies carried out on peripheral target tissues such as rat uterus and chick oviduct although the general principles appear to be applicable to all target cells. For further details see [2] in Chapter 1 and V. H. James & J. R. Pasqualini (eds) (1980) *Hormonal Steroids* (Pergamon Press).

Sex steroids are transported to their target tissues in the blood and over 90 per cent of circulating steroids will be bound to specific plasma proteins called globulins. Oestrogen and testosterone bind to a specific sex hormone binding globulin whereas progesterone binds to the cortisol binding globulin; all steroids can also bind to albumin, though with a lower affinity. An equilibrium exists between the amount of free and bound circulating steroids, but it is only the free steroid which is biologically active. Thus the binding of steroids allows for a large potential 'reserve' of circulating free hormone and is also important for protection against rapid metabolism of the hormones. High concentrations of a further protein which specifically binds oestrogens (alpha-fetoprotein) is found in the fetus of female animals and humans and it has been suggested that this may function to protect the fetus against excessive hormonal stimulation from the high levels of maternal oestrogen. This may be particularly important at the time of sexual differentiation when levels of oestrogen (formed from testosterone) may be critical (see Chapter 11 Section 3(ii)).

The synthesis of highly specific ^{3}H-labelled steroids in the late 1950s and early 1960s was a major breakthrough in identifying and quantifying steroid-receptor interactions and since that time autoradiographic techniques, procedures for fractionation of target tissues and biochemical studies on cell-free systems has greatly advanced our knowledge. More recently the production of specific receptor-antibodies to oestrogen and progesterone and the application of genetic engineering techniques has been of value in understanding how steroids exert their effects (see Chapter 1 Section 11). It is well established that cells in the hypothalamus, the preoptic area, pituitary gland and amygdala retain ^{3}H-labelled sex steroids (oestrogen, progesterone and testosterone), the highest concentrations being located in the hypothalamus and pituitary gland. More recently oestrogen receptors have been located in the locus coerulus which is of interest in view of the fact that noradrenergic fibres terminating in the hypothalamus arise from cell bodies located in this area. Brain regions which retain labelled sex steroid are also known to be involved in the control of gonadotrophin secretion and are responsive to steroid implants.

In the brain and pituitary, oestradiol and progesterone can interact directly with their receptors without further metabolism, although testosterone is subject to metabolic transformations in neuroendocrine tissue, and the metabolites interact with the hormone receptors. One of the important reduction products of testosterone is 5α-dihydrotestosterone (5α-DHT) which has been implicated in the feedback regulation of gonadotrophin secretion. Alternatively, testosterone which can be converted into oestradiol may mediate androgen-facilitation of sexual behaviour and sexual differentiation of the brain (see Chapter 11 Section 3). Progesterone is also reduced to 5α-dihydroprogesterone in the brain and pituitary and these metabolites may be important in the progesterone effects on sexual behaviour and ovulation. Metabolites of oestrogen which have provoked some interest are the catecholoestrogens (2- and 4-hydroxylated oestrogens) since there may be links between these steroids and catecholamines. Catecholoestrogens are formed in distinct brain areas such as the hypothalamus and amygdala, but not in the pituitary, and there is evidence for their participation in neuroendocrine functions. Actions of catecholoestrogens which have been described include inhibition of enzymes in the synthesis and degradation of catecholamines, effects on the turnover rates of catecholamines by competing for monoamine oxidase enzymes which metabolise the catecholamines, binding to hypothalamic oestrogen receptors, and stimulation of adenylate cyclase. Such observations have suggested that catecholoestrogens may act on aminergic systems although their physiological role in feedback mechanisms is not well established. The steroid transformations, described above, which occur in the brain and/or pituitary tissue of the rat are outlined in Figure 8.5.

It is now generally recognised that the major effect of steroid hormones is to increase protein synthesis by stimulating DNA-dependent messenger RNA (mRNA) synthesis and ribosomal RNA (rRNA) synthesis. Stimulation of mRNA is likely to be the rate-limiting step of steroid action and it is thought that steroids exert their effects at the transcriptional level, i.e. render sequences of the DNA molecule accessible for transcription by RNA polymerase. If this generalised model for steroid action within target cells is applicable for feedback effects of sex steroids, one would need to know how this is translated into an alteration of gonadotrophin secretion. Does the uptake of steroids by target neurones alter the synthesis and metabolism of Gn-RH, the turnover of neurotransmitters, the excitability of neurones, the

Figure 8.5: Some Important Steroid Transformations which Occur in the Rat Brain and/or Pituitary, prior to Binding with their Receptors.

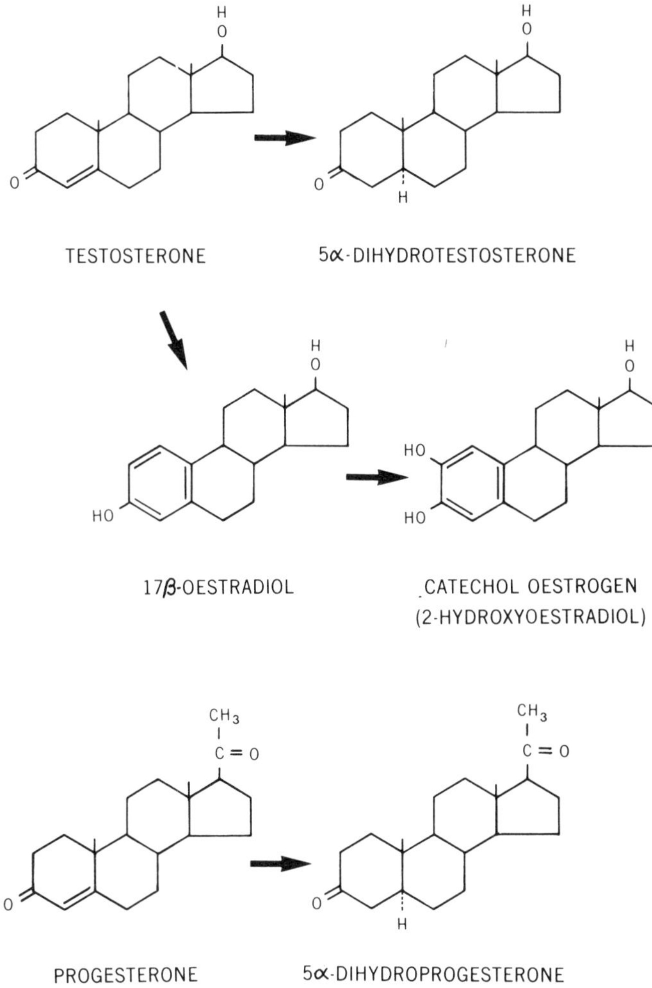

formation of synapses? Although evidence suggests that oestrogen can exert all these effects on the central nervous system the molecular events which link steroid-receptor activation to a change in gonadotrophin secretion remain a mystery and is one of the key questions pertaining to mechanisms controlling anterior pituitary secretions.

The idea that central steroid receptor interactions are biologically important is supported by evidence obtained from steroid implants into neural tissue and the use of steroid antagonists which can induce functional changes in the hypothalamic-pituitary-gonadal axis. Genomic effects of steroids have been implicated by alterations in RNA synthesis in the pituitary gland following oestrogen administration. Less consistent results of steroid induced increased RNA or protein synthesis have been obtained from studies on hypothalamic steroid receptors, although some steroid effects (e.g. oestrogen induction of sexual behaviour and progesterone facilitation of LH secretion in the rat) can be blocked by protein synthesis inhibitors. Other approaches to investigating gonadal steroid effects at brain level have indicated more acute mechanisms of action than the genomic effects. For example, oestrogen has been shown to increase the activity of peptidase enzymes that break down Gn-RH whereas it has opposite effects on monoamine oxidase and tyrosine hydroxylase activity, enzymes which metabolise and synthesise catecholamine neurotransmitter respectively. Furthermore, electrophysiological investigations have suggested that steroids may have a direct membrane effect on neurones thereby altering neuronal excitability, and finally studies with hypothalamic synaptosomes (see Chapter 2 Section 5(i)) have suggested that steroid hormones may exhibit feedback effects on Gn-RH release, for example, in which nuclear genomic mechanisms and protein synthesis cannot be activated to mediate such effects.

8:6 Neural Regulation of Gonadotrophin Secretion

(i) *Neural Inputs to the Gn-RH Neurosecretory Cells*

Neural inputs to areas of the brain containing Gn-RH immunoreactive material are diverse and exemplify the highly complex integration of neural signals, generated not only by steroid feedback mechanism but also by environmental factors and non-specific stimuli such as emotional stress. The hypothalamic dopaminergic neurones are important in modulating the secretion of Gn-RH. Axo-axonic connections have been shown to exist between these dopaminergic nerve terminals and Gn-RH containing nerve endings in the median eminence, and it is clear that dopamine has complex effects on Gn-RH release and gonadotrophin secretion (see below). Facilitatory connections with the hypothalamus are thought to arise from noradrenergic neurones located in the locus coerulus of the brain stem while mid-brain 5-

hydroxytryptamine neurones, which have a significant projection to the suprachiasmatic nucleus (see Chapter 9 Section 2), have been implicated in generating the neural signal for the LH surge. The mid-brain, forebrain and limbic-system provide significant inputs to the hypothalamus which probably subserve the effects of external (visual, auditory and olfactory) and internal factors on Gn-RH release. The limbic system conveys the olfactory signals, which are important in reproductive behaviour in lower mammals, while light signals could affect gonadotrophin secretion via the direct retino-hypothalamic pathway.

In the rat, stimulation and lesions studies have shown that neural inputs from the rostral hypothalamus and preoptic area to the basal hypothalamus are essential for the positive feedback effects of oestrogen and the initiation of the preovulatory LH surge, while the isolated basal hypothalamus (hypothalamic 'island') is in itself, capable of maintaining normal, low level secretion of the gonadotrophins. Conversely, in monkeys complete deafferentation of the basal hypothalamus does not disrupt spontaneous or oestrogen induced LH surge although different results have been obtained from different laboratories. Recent work with primates has shown that, following destruction of the Gn-RH neurosecretory cells by radiofrequency lesions of the arcuate nucleus, oestrogen administration in ovariectomised animals can induce both a negative and a positive feedback effect on LH secretion from the pituitary provided the animals were infused with hourly pulses of Gn-RH for several days prior to and during oestrogen treatment[7]. Such studies were interpreted as indicative of an ovarian signal, as opposed to a neural signal, which stimulates the preovulatory surge of LH and that Gn-RH simply plays a permissive role in maintaining the integrity of the pituitary gonadotrophs. It may well be that in rats the entrainment of ovulation to light/dark cycles (see Chapter 9 Section 3(ii)) exemplifies the importance of neural signals from the external environment, whereas higher mammals show less dependency on external factors. Nevertheless the growing body of evidence, both in primates and other mammals, implicating the importance of the frequency and amplitude of Gn-RH secretory pulses in determining the pituitary responses suggest a strong neural component in the central control of gonadotrophin secretion. Whether this neural component is initiated by ovarian signals in terms of oestrogen feedback (a pelvic clock in the timing of ovulation) or by exteroceptive factors and circadian rhythms (a neural clock) remains to be established, although in animals such as the rat or seasonal breeders (see Chapter

9 Section 3) exteroceptive signals clearly play a much greater role in determining ovulatory cycles than in primates, including man.

(ii) *Electrophysiological Correlates of Gonadotrophin Secretion*
Early experiments by Cross and Green in 1959 demonstrated that the spontaneous rate of neuronal firing in the hypothalamus of female rats was highest just prior to ovulation compared to other stages of the oestrous cycle. Since that time advances in electrophysiological recording techniques have provided insights into possible neural connections of parvicellular neurosecretory cells, their putative neurotransmitters and the influence of sex steroids on neuronal excitability. However, the inability to demonstrate conclusively that a particular neurone is neurosecretory, or to identify what hypothalamic hormone is secreted by that cell, has limited the contribution of electrophysiological studies in elucidating neural control of adenohypophyseal function. Antidromic stimulation techniques have identified a wide distribution of cells in the basal hypothalamus and more rostral regions in the anterior hypothalamus and preoptic areas that project to the median eminence. It has also been shown that some tuberoinfundibular neurones can be antidromically activated not only by stimulation of the median eminence but also by stimulation of the amygdala, thalamus, paraventricular nucleus, anterior hypothalamus and preoptic area (see [12] in Chapter 2). This suggests that axon collaterals of the neurosecretory cells may also make extrahypothalamic reciprocal connections with brain areas known to provide a functional input to the hypothalamus (see Figure 8.6).

The effect of gonadal steroids on the electrical activity of hypothalamic neurones has been investigated either by injecting steroids peripherally or by their direct application to neurones by microiontophoresis (see Chapter 2 Section 3(iii)); both excitatory and inhibitory responses have been reported. Similar findings were obtained when LH, FSH or Gn-RH were microiontophoretically applied to neurones in the basal hypothalamus, although inhibitory responses predominated. The responses of hypothalamic neurosecretory cells to a wide spectrum of putative neurotransmitters show no consistent pattern but generally both dopamine and noradrenaline inhibit the firing rate of these units. Until immunohistochemical identification of peptidergic tuberoinfundibular neurones can be correlated with electrophysiological recordings and hormone secretion, electrophysiological experiments will, at best, be simply indicative of neural events controlling adenohypophyseal secretions.

Figure 8.6: Proposed Connections of Hypothalamic Neurosecretory Cells in the Basal Hypothalamus Based on Evidence from Electrophysiological and Morphological Studies. The diagram indicates axo-axonic connections in the median eminence, recurrent collaterals (see Chapter 2 Section 3(iii)) of the neurosecretory cells and reciprocal connections between extrahypothalamic brain areas (For explanation see text). AHA – anterior hypothalamus; DM – dorsomedial; POA – preoptic area; PV – paraventricular nucleus.

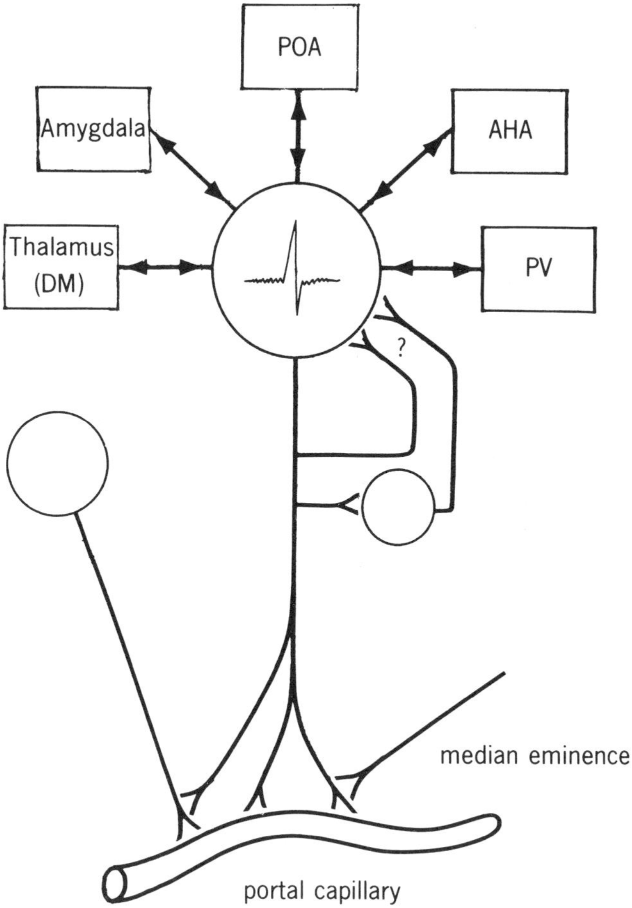

(iii) *Putative Neurotransmitters*

A wide spectrum of pharmacological studies have shown that several neurotransmitter candidates are involved in the regulation of gonadotrophin secretion. The catecholamines, dopamine and noradrenaline and the indolamine, 5-hydroxytryptamine (5-HT), have probably received the most attention although results of studies which have set out to manipulate brain amine levels, either by drugs which deplete the amines and/or specific agonist and antagonists, often produce conflicting results. Dopamine has been shown to both stimulate and inhibit gonadotrophin release although it is generally agreed that the effects of dopamine are primarily inhibitory to LH and FSH secretion. Recent research has implicated the existence of both stimulatory and inhibitory dopamine receptors controlling Gn-RH release and other studies suggest that while the dopaminergic tuberoinfundibular tract may tonically inhibit Gn-RH release, another hypothalamic dopamine pathway, the incerto-hypothalamic tract may be stimulatory. Noradrenaline stimulates gonadotrophin release and it appears that it is particularly important in the positive feedback effects of oestrogen. However, it should be noted that destruction of noradrenergic nerve terminals only temporarily disrupts ovarian cyclicity suggesting that other neural mechanisms can compensate for the loss of a noradrenergic input to the hypothalamus. The turnover rates of both hypothalamic dopamine and noradrenaline have been correlated with the differential effects of oestrogen feedback. During the pre-ovulatory oestrogen surge, dopamine turnover and the concentration of dopamine in the hypophyseal portal system is reduced, whereas noradrenaline turnover is increased. These observations lend further credence to the evidence for inhibitory and stimulatory roles of dopamine and noradrenaline respectively on Gn-RH secretion. The precise role of 5-HT in gonadotrophin release is uncertain although in an attempt to explain a wealth of conflicting literature it has been suggested that the effects of this neurotransmitter are closely related to circadian rhythms since 5-HT seems to be important in the regulation of these rhythms. Thus drug treatments which result in a prolonged elevation or decrease in brain 5-HT tend to inhibit gonadotrophin release whereas a rapid change in brain 5-HT levels (be it raising or lowering) causes a temporary stimulation. Other neurotransmitters have been implicated in stimulating gonadotrophin secretion and these include acetylcholine (by acting on muscarinic receptors), histamine and prostaglandins.

(iv) *Neuropeptides and Gonadotrophin Secretion*

Several peptides which are present in the hypothalamus have been shown to alter gonadotrophin secretion – the opioid peptides, vasoactive intestinal polypeptide, neurotensin, substance P, cholecystokinin, gastrin – and most of the experimental evidence points to a hypothalamic site of action, which would agree with the evidence for a neurotransmitter/neuromodulatory role of CNS peptides (see Chapter 10 Section 3). The opioid peptides, the enkephalins and β-endorphin inhibit gonadotrophin release, and it has been suggested that this effect is mediated by dopamine since the opioid peptides and morphine increase the turnover and release of hypothalamic dopamine. Substance P, neurotensin and VIP stimulate LH and FSH secretion by acting on the hypothalamus, whereas the effects of cholecystokinin and gastrin are inhibitory. Despite the growing interest in the effects of peptides on anterior pituitary function as yet no definite physiological role has been ascribed and it may be that if, indeed, their effect is one of 'fine-tuning' of the system small changes which occur in relation to regulation of reproductive cycles may be extremely difficult to detect experimentally.

References

1. Adler, N. T. (ed.) (1981) *Neuroendocrinology of Reproduction: Physiology and Behaviour*. Plenum Publishing Company.
2. Austin, C. R. & Short, R. V. (eds) (1979) *Mechanism of Hormone Action*. Book 7: *Reproduction in Mammals* (Cambridge University Press, Cambridge).
3. Catt, K. J., Harwood, J. P., Clayton, R. N., Davies, T. F., Chan, V., Katikineni, K., Nozu, K. & Dufau, M. L. (1980) Regulation of Peptide Hormone Receptors and Gonadal Steroidogenesis. *Recent Progress in Hormone Research*, **36**: 557-622.
4. Chan, L., Means, A. R. & O'Malley, B. W. (1978) Steroid Hormone Regulation of Specific Gene Expression. *Vitamins and Hormones*, **36**: 259-96.
5. Franchimont, P., Verstraelen-Proyard, J., Hazea-Hagelstein, M. T., Renard, C., Demoulin, A., Bourguignon, J. P. & Hustin, J. (1979) Inhibin: from Concept to Reality. *Vitamins and Hormones*, **37**: 244-302.
6. Gorski, R. A. (1979) The Neuroendocrinology of Reproduction: an Overview. *Biology of Reproduction*, **20**: 111-27.
7. Knobil, E. (1980) The Neuroendocrine Control of the Menstrual Cycle. *Recent Progress in Hormone Research*, **36**: 53-88.
8. Krieger, D. T. & Hughes, J. C. (1980) Various Articles on Neuroendocrine and Gonadal Functions, Section V in D. T. Krieger & J. C. Hughes (eds) *Neuroendocrinology – a Hospital Practice Book* (Sinauer Assoc. Inc., Sunderland, MA).
9. Lincoln, G. A., (1979) Pituitary control of the testis. *British Medical Bulletin*, **35**: 167-72.
10. McCann, S. M. (1975) Regulation of Follicle Stimulating Hormone and Luteinizing Hormone, in *Handbook of Physiology and Endocrinology*, IV, part 2, pp. 489-517.
11. Millar, R. P. (ed.) (1981) Various articles on the Biosynthesis and Actions of Luteinizing Hormone Releasing Hormone (Gn-RH) in *Neuropeptides: Biochemical and Physiological Studies* (Churchill Livingstone, Edinburgh).

12. Ojeda, S. R. (1980) The Endocrinology of Sexual Development (Symposium) Series of articles in *Federation Proceedings*, **39**: 2353-441.
13. Sanborn, B. M., Heindel, J. J. & Robison, A. J. (1980) The Role of Cyclic Nucleotides in Reproductive Processes. *Annual Review of Physiology*, 42: 37-58.
14. Sharp, P. J. & Fraser, H. M. (1978) Control of Reproduction, in S. L. Jeffcoate & J. S. M. Hutchinson (eds) *The Endocrine Hypothalamus* (Academic Press, New York), pp. 271-332.
15. Tyson, J. E. (ed.) (1978) Various articles on Reproductive Endocrinology in *Clinics in Obstetrics and Gynaecology*. Vol. 5, (2).

9

ENDOCRINE RHYTHMS

There is a growing interest of both neuroscientists and chronobiologists in the neural mechanisms controlling biological rhythms, and with respect to the neuroendocrine system, the anatomical localisation and biochemical characteristics of pacemakers, or biological clocks, are being actively researched. A discussion on rhythms in neuroendocrine control systems embraces a wide spectrum of hormonal secretory patterns ranging from the typical pulsatile (episodic) secretion to circadian oscillations to cyclical and seasonal variations. In the present chapter a few examples of these 'rhythms' will be briefly described to illustrate some important aspects of temporal patterns of hormone release and their postulated controlling mechanisms.

9:1 Pulsatile or Episodic Hormone Secretion

It is generally considered that most, if not all pituitary hormones and hypothalamic hormones are secreted in a pulsatile pattern and that this is due to a 'drive' mechanism in the central nervous system which is independent of stimuli and feedback mechanisms. Such a phenomenon has been shown in the secretion of ACTH and cortisol, the gonadotrophins, TSH, growth hormone and prolactin and has been studied in both animals and humans with indwelling catheters placed in an appropriate vein, so that blood samples may be collected at frequent intervals over a 24 hour period without inducing stress. For example, growth hormone (GH) secretion shows marked spontaneous surges throughout the day, peaks occurring every 3-4 hours falling abruptly to undetectable levels between each pulse (see Figure 7.3). Similarly the secretion of other hormones occurs in erratic pulses although the frequency and amplitude of the pulses vary between hormones and may be modified by both extero- and interoceptive factors, e.g. light, sleep, feedback mechanisms. A further interesting example of episodic hormone secretion is the release of oxytocin in lactating nursing rats. This occurs in 'bursts' with a periodicity of 4-8 minutes even though the stimulus for secretion – suckling of the pups – is continuous.

The pattern of luteinising hormone (LH) and gonadotrophin releasing hormone (Gn-RH) secretion is also pulsatile and recently much interest has been directed toward the qualitative pattern of Gn-RH secretion. It has been proposed that the actual 'drip' rate of Gn-RH may be the important determinant for regulating pituitary responsiveness and that the frequency of Gn-RH pulses is determined by the integration of intero- and exteroceptive factors (see Chapter 8 Sections 3(iii) and 4). Continuous infusion of Gn-RH in female monkeys leads to abolition of the pituitary response, whereas intermittent infusion at about one pulse per hour re-establishes normal pituitary function. Infusions at a higher frequency also inhibits gonadotrophin secretion suggesting that the pattern of Gn-RH release is of paramount importance in determining the responsiveness of the pituitary (see [7] in Chapter 8). In addition, it is well established that Gn-RH stimulates the synthesis of pituitary gonadotrophins and it has been proposed that the frequency and amplitude of Gn-RH pulses can alter the sizes of the readily releasable pool and the reserve pool of pituitary LH. Thus a dynamic situation between the synthesis and secretion rates and the reserve of LH could exist, this being determined by the pattern of Gn-RH secretion and feedback effects of gonadal steroids. The same principles may apply for all other hypothalamic hormones and circumstantial evidence suggests this is true for the control of TSH secretion.

The physiological significance of pulsatile hormone release has been attributed to desensitisation or 'down regulation' of hormone receptors (see Chapter 1 Section 10). This can occur if the target cells are continually challenged by a particular hormone, but intermittent hormone secretion would serve to protect against desensitisation. However, the neural mechanisms controlling pulsatile secretion are not well understood although in the case of pituitary hormones which are regulated by two hypothalamic releasing hormones (GH and prolactin) the interplay between stimulatory and inhibitory factors on the pituitary may, in part, explain some patterns of hormone release (see Figure 7.3).

9:2 Circadian Rhythms

(i) *Endogenous Endocrine Mechanisms*

Circadian rhythms are endogenous oscillations with a period of approximately 24 hours. A wide spectrum of physiological functions, ranging from locomotor activity to metabolic fluctuations, demonstrate circadian rhythms and generally the light/dark cycle is the major entraining stimulus. However, the light/dark cycle does not dictate the circadian rhythms since when animals or man are exposed to constant light or dark conditions the rhythms still persist with a periodicity of approximately 24 hours although they become 'free running'. Frequently the terms diurnal or nocturnal are used interchangeably with circadian but strictly speaking diurnal and nocturnal rhythms more accurately describe 24 h cycles whose endogenous nature is in doubt i.e. periodicities resulting from environmental fluctuations or stimulations which disappear when light or other conditions are kept constant. Hormone secretions show circadian rhythms but many are more strongly entrained to the sleep/wake cycle rather than the light/dark cycle and thus nocturnal animals (active at night, e.g. rat) have endogenous rhythms approximately 12 hours out of phase with diurnal animals (active during the day, e.g. man usually).

The majority of the anterior pituitary hormones show circadian rhythms in their pattern of secretion, this being superimposed on their pulsatile release. Some of these circadian rhythms are marked and well documented, particularly for ACTH and cortisol in which the blood levels of both hormones in humans show peak levels in the early morning (see Figure 5.4) while peak levels of ACTH and corticosterone in the rat occur in the evening. Peak levels of human TSH are considered to occur in the evening but the relationship between circadian changes of TSH and thyroid hormone secretion are not as clear-cut as in the pituitary-adrenal axis. Circadian changes in the secretion of the gonadotrophins are very complex and show species variation (see section 3 and Chapter 8 section 6). Hypothalamic concentrations of CRF, TRH, and Gn-RH also appear to exhibit circadian rhythms (see [11]) implicating that the endogenous rhythms of pituitary hormone secretions are neurally mediated.

In contrast to the above-mentioned hormones, the secretion rates of GH and prolactin are more directly controlled by factors such as sleep and their rhythms of secretion are less related to circadian light-entrained rhythms. The release of GH and prolactin occurs in infrequent pulses throughout the awake period but an increase in the frequency of pulses occurs following the onset of sleep. Increased prolactin secretion can occur throughout the sleep period but the major episode of growth hormone release usually occurs during the early onset of sleep (see Figure 7.3). These rhythms are particularly marked in children and young animals and are totally dependent on the onset of sleep since it does not occur when sleep is delayed or prevented. There is also evidence that LH secretion during puberty is particularly associated with sleep patterns although this is not found in the adult (see [1, 13]).

In summary, circadian rhythms of hormone secretion entrained to the light/dark cycle are widespread but exogenous factors may sometimes override the natural rhythm. The circadian secretions occur as a result of neurally directed changes in the frequency and amplitude of the episodic release of the hormone. Transitory changes in light conditions or short-term delay or absence of sleep will not prevent the normal endogenous rhythms occurring but, if there is a persistent reciprocal change in the light/dark cycle, the circadian rhythm will gradually shift over a period of several weeks to a newly entrained rhythm. For further reading see [5, 6, 9, 10, 11].

(ii) *The Pineal Gland*

The release of the hormone, melatonin from the pineal gland also shows a marked circadian rhythm entrained to the light/dark cycle (see Figure 9.1). The pineal gland operates as a neuroendocrine transducer (see Chapter 1 Section 5) and is innervated by noradrenergic post-ganglionic fibres of the superior cervical ganglia which stimulates the conversion of 5-hydroxytryptamine into melatonin in the pinealocytes. *N*-acetyltransferase is the enzyme controlling the rate-limiting step in the production of melatonin and a nocturnal peak in the activity of *N*-acetyltransferase and melatonin production has been observed. The pineal in lower vertebrates is itself photosensitive but in mammals neural pathways from the retina to the pineal (some via the suprachiasmatic nucleus (see Figure 9.2)) are considered important in controlling its photoperiodicity. The function of the pineal gland in mammals is not understood although melatonin can inhibit gonadal activity and may play some role in regulating the photoperiodic control of seasonal breeding in a few species e.g. hamster (see Section 3). For further reading on the pineal gland, its neuroendocrine regulation, secretion of melatonin, and its relation to circadian rhythms see [2, 4, 6, 13]. The marked circadian rhythms observed in both corticosteroid secretion and pineal *N*-acetyltransferase activity have provided good experiment models in which the mechanisms controlling light-entrained

Figure 9.1: Diagram Showing the β-adrenergic Control of Melatonin Synthesis in the Pineal Gland. Stimulation of noradrenergic nerve terminals (NA) increases the activity of *N*-acetyltransferase (N-AcT) via adenyl cyclase (AC) activation and an increase in cyclic AMP (cAMP) synthesis. The circadian rhythms in the concentrations of 5-hydroxytryptamine (5-HT) and melatonin and the activity of N-Act are also shown.

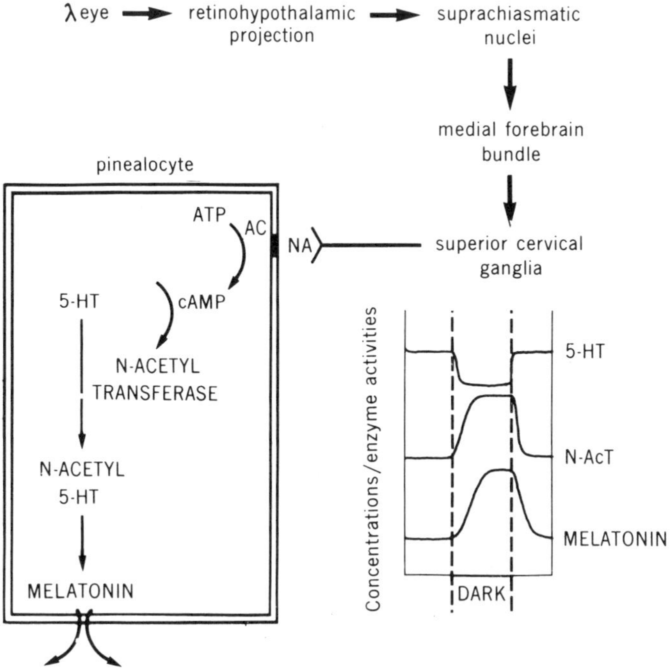

circadian rhythms may be investigated.

(iii) *The Suprachiasmatic Nucleus*

Many biological circadian rhythms, including endocrine rhythms, have been described and recently much attention has been given to the precise localisation of the neural 'biological clock' mechanisms for regulating such rhythms. During the last few years independent lines of research have focused attention on the suprachiasmatic nuclei (SCN) in the anterior hypothalamus as a possible neural centre controlling circadian rhythms. In rodents the SCN are a pair of well-defined ovoid nuclei containing small densely packed neurones (see Figure 1.2), although in higher mammals these structures are either poorly defined or absent. Three major afferent pathways to the SCN have been described: (1) a pathway which originates in the retina and

Figure 9.2: Main Neural Connections of the Rat Suprachiasmatic Nuclei (SCN).

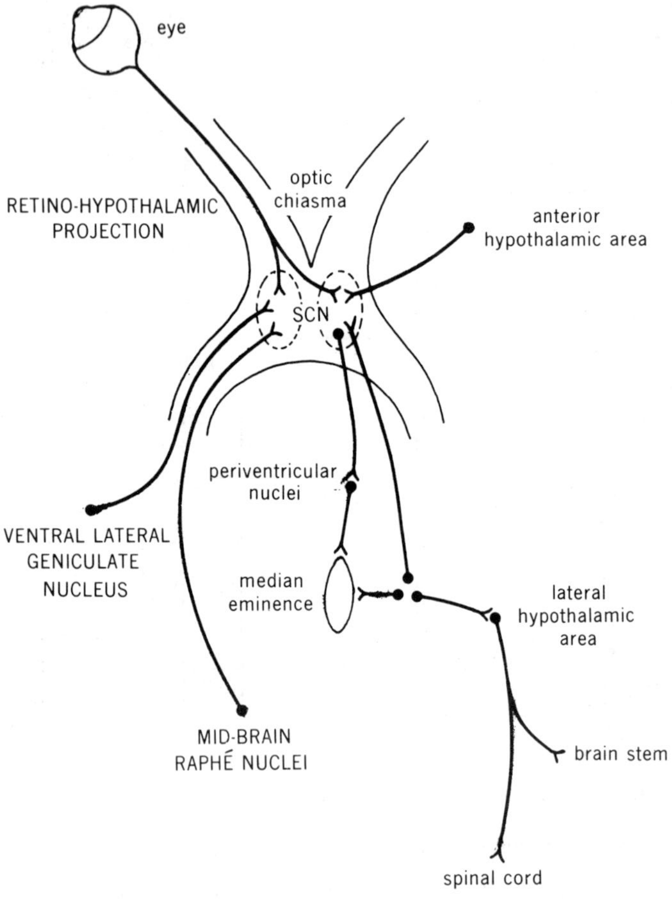

reaches the SCN via the optic nerves and optic chiasm – the retino-hypothalamic tract; (2) fibres from the ventral lateral geniculate nuclei which reaches the SCN either via the medial lemniscus or via the optic tract and (3) a 5-hydroxytryptamine (5-HT) projection from the mid-brain raphe nuclei. Inputs to the SCN are therefore linked firstly with light/dark cycles from retinal projections, which is consistent with the entrainment of circadian rhythms to these cycles, and secondly with sleep cycles from the raphe nuclei, which is consistent with the known involvement of 5-HT pathways from these brain stem nuclei in regulating sleep patterns.

Very little is known about the fine efferent fibres originating in the SCN although there is evidence for projections to ventromedial, dorso-medial and arcuate nuclei and to the median eminence. Some of the major connections of the SCN are diagrammatically illustrated in Figure 9.2.

Lesions of the SCN in rodents abolish the circadian rhythms of ACTH/cortisol, pineal melatonin secretion and various rhythms related to reproductive processes (see Section 3), although lesions of other brain areas such as the ventromedial and dorsomedial nuclei can also affect inherent fluctuations of hormone secretion. Interestingly the SCN shows a circadian rhythm in metabolic activity as indicated by glucose uptake; this is not observed in adjacent brain areas. More direct evidence that the SCN regulates circadian rhythms has been provided by electrophysiological studies in which a clear 24h rhythm of neural activity was recorded from neurones in the SCN, and this rhythm persisted even in Halasz hypothalamic 'island' preparations (see Chapter 2 Section 3(ii)) containing the SCN.[12] Since the SCN have been implicated in the regulation of reproductive cycles, there has been considerable interest in establishing whether the suprachias-matic region contains neurosecretory neurones. Vasopressin and its carrier neurophysin has been located in the rat and mouse SCN and there is some evidence for the presence of Gn-RH, somatostatin, neurotensin and melatonin, and perhaps other peptides in the SCN. However, it is generally thought that circadian rhythms of pituitary hormones are generated indirectly via inputs from the SCN and/or other brain regions to the appropriate neurosecretory cells.

Most experimental evidence which supports the idea that the supra-chiasmatic nuclei are the location of a biological clock in mammals is derived from investigations with small rodents, particularly rats and hamsters although more recent studies have suggested that this region of the brain may be important for generating circadian rhythms in many, if not all, vertebrate classes. In higher mammals, such as primates, a more complex regulation of circadian rhythms is apparent where a greater capacity for dissociation of the endogenous circadian rhythm is present. However, the evidence that the SCN is an important generator of light/dark entrained circadian rhythms is compelling since ablation of the nuclei abolishes these rhythms and the only retinal projection which results in a loss of circadian rhythmicity is the retinohypothalamic tract terminating almost exclusively in the suprachiasmatic nuclei.

It has now become apparent that hormones themselves can influence circadian pacemakers most likely via direct actions on central 'biological clock' regions. For example the daily onset of locomotor activity in hamsters fluctuates during the oestrous cycle and this is dependent on oestrogen secretion, while corticosterone can entrain a number of biological rhythms. Thus hormones which display circadian oscillations in their secretion rates also influence other daily rhythms. In some vertebrate classes the pineal complex may function as a biological clock, although in mammals removal of the pineal does not interfere with normal circadian rhythms. The pronounced effects of the pineal on reproductive rhythms of photoperiodic mammals are more readily attributable to pineal anti-gonadal activity than to pineal-induced changes in circadian rhythms (see Section 3). In summary, the SCN seems to be the most likely neural centre for the regulation of light/dark entrained circadian rhythms observed in adenohypophyseal hormone secretions although hormonal effects and inputs from other brain regions can clearly influence such endogenous oscillations. For further reading on circadian rhythm control see [2, 5, 6, 9, 10].

9:3 Reproductive Cycles

The reproductive cycles of mammalian species present a plethora of information regarding rhythms of hormone secretion – pulsatile patterns imposed on circadian rhythms which in turn can influence not only continuous ovulatory cycles of finite duration but also the photoperiodic control of seasonal breeding. In Chapter 8 we considered some of the ways in which reproductive hormones interact but only on an internal basis. In this section the coordination of reproductive cycles with environmental factors will be discussed, particularly the relationship between the light/dark (LD) entrained circadian rhythms, ovulation and sexual receptivity.

(i) *Reproductive Cycles and the Environment*

The period of sexual receptivity in mammals is defined as oestrus and it is of obvious advantage if ovulation and oestrus coincides with the appearance of a suitable male. Some animals, for example cats, rabbits and ferrets, ovulate in response to mating. These mammals are known as 'reflex' or induced ovulators and ovulation is dependent upon sensory stimulation of the cervix and genital areas. Such animals stay in a state of almost continuous oestrus which, in the case of the ferret, can be from early Spring until September or in the domesticated rabbit may be almost permanent. In these animals mating induces a reflex surge of LH secretion and ovulation occurs about ten hours later. All other species are classified as spontaneous ovulators and have defined reproductive cycles in which ovulation is not dependent on mating. The duration of the cycle varies with the species (see Table 9.1) with a functional significance between cycle length and life expectancy of any one species; short cycles promoting breeding and facilitating production of multiple litters during any one breeding season. Mammals that have many oestrous cycles are said to be polyoestrous, some of which cycle continuously throughout the year (e.g. women, the laboratory rat, mouse and the guinea pig) while others have a series of sequential cycles throughout a limited breeding season such as sheep and many deer. Then there are dioestrous mammals having just two heats a year (such as the bitch) and monoestrous ones with a single heat (e.g. silver fox and some martens).

In most cases breeding seasons are correlated with changes in the environment so that the birth of the young occurs at a time when temperature and food availability are advantageous to their survival. For example the autumn mating of ewes will give lambing during January and February and allow the lambs to be weaned on spring grass. The ewes will come into oestrus as the length of daylight decreases during the autumn. In some species photoperiods may not only determine the onset of oestrus but may also determine the implantation of the fertilised blastocyst. For instance, mating and fertilisation takes place but the blastocyst may not implant immediately, remaining free in the uterus until changes in the day-length prompt implantation. In this way animals may mate in the spring, delay implantation to the autumn and give birth the following spring. Alternatively animals with shorter gestation may mate in the autumn and delay implantation for a certain period with the result that the young are born in the spring. Delayed implantation can be terminated by changing the photoperiod,

Table 9.1:

Species	Oestrous cycles	Duration of cycles (days)	Average gestation period (days)
Laboratory rat	Polyoestrous (continuous)	4–5	21
Laboratory mouse		5	19
Guinea pig		16	68
Ferret	None (reflex ovulators)	–	42
Domestic rabbit		–	31
Domestic pig	Polyoestrous (continuous)	21	113
Domestic cattle	Polyoestrous (almost continuous)	21	282
Domestic sheep	Polyoestrous (seasonal)	16–17	150
Indian elephant	Polyoestrous		623
Rhesus monkey	Polyoestrous (continuous)	28	164
Chimpanzee		34	237
Woman		28	280

but the mechanisms controlling implantation are unknown.

Having defined certain temporal variations of reproductive cycles and breeding seasons let us now consider the relationship between the circadian system and reproductive system – firstly in those species which have ovulatory cycles throughout the year and secondly in mammals who are seasonal breeders and whose onset of oestrus is determined by the changing length of day.

(ii) *Non-seasonal Breeders*

It is generally accepted that in many species of so-called 'non-seasonal breeders' a neural signal generated at circadian intervals determines the timing of ovulation. Some animals when maintained in constant light either cease to cycle or have lengthened oestrous cycles, and the timing of ovulation can be shifted by twelve hours by reversing the light/dark cycle. Light-entrained LH surges can be observed in the oestrogen treated ('primed') ovariectomised rat and hampster and the precise timing of the neural signal for the preovulatory LH surge in these animals is tightly linked with the light/dark cycle. Conversely, there is no evidence for a circadian organisation of reproductive cycles in guinea pig and monkey, and it has been suggested that in these species steroid feedback may be sufficient to the timing of ovulation and organisation of the oestrous cycle. This of course raises the same question of a neural versus a pelvic clock (see Chapter 8 Section 6), and although it may be somewhat premature in light of the current evidence, a few tentative generalisations on circadian pacemakers and oestrous cycles may be made.

It is accepted that in rats and hamsters the anterior hypothalamus/preoptic area is important for generating the neural trigger for ovulation though some reports challenge the concept that the preoptic area is indispensable for ovulation, particularly in the long term. On the other hand it seems that only the basal hypothalamic-pituitary axis is required for ovulation in monkey and guinea pigs; animals which show no evidence of internal circadian oscillators governing the timing of ovulation. The circadian pacemaker in rats and hamsters appears to reside in the suprachiasmatic nuclei (see Chapter 9 Section 2). Lesions in the SCN disrupt oestrous cycles, can inhibit the daily surges of LH and prolactin in the ovariectomised oestrogen primed rat and prevent the diurnal and nocturnal rise of prolactin secretion in pseudopregnant rats. However, the close proximity of the SCN and the preoptic area and the inherent difficulty in making discrete lesions confined to the small SCN have led to conflicting conclusions concerning the association between a circadian oscillator in the SCN, the preoptic area and the neural timing of ovulation.

Apart from the SCN, the pineal gland has also been implicated as a circadian pacemaker determining the timing of ovulation. Numerous experiments, particularly in rodents, have shown that melatonin has anti-gonadal properties and it is likely that any effects of the pineal on reproductive cycles are mediated by an action of melatonin. The ovary, uterus and pituitary gland have been suggested as sites of melatonin's anti-gonadal activity and there is also evidence for a site in the preoptic and suprachiasmatic regions. Melatonin is specifically bound in the rat hypothalamus and endogenous melatonin has been localised in the SCN by fluorescent immunohistochemistry. Thus melatonin may influence the circadian pacemaker of the SCN, thereby affecting reproductive rhythms. In summary, the timing of ovulation in some non-seasonal breeders is entrained to light/dark cycles and circadian rhythms and this entrainment probably involves the SCN. On the other hand, oestrous cycles in some mammals appear to be independent of autonomous circadian pacemakers and ovulation is determined solely by the feedback of gonadal steroids.

(iii) *Seasonal Breeders*

It has long been recognised that the seasonal patterns of reproduction are controlled by the environmental photoperiod and the onset of the breeding season is normally connected with changes in the day length. In many species, females will only show cyclic ovarian activity for limited periods of the year while the males of the species only develop active testes at this correspoding period. Animals whose breeding season is stimulated when the days shorten, e.g. sheep and deer, are termed 'short day breeders' while other mammals, e.g. horse, come into season with the lengthening days and are sometimes referred to as 'long day breeders'. Just how animals compare the day length over several days or weeks and how retinal stimulation is converted into a neural signal which ultimately increases gonadotrophin secretion is a subject of much debate. However, in recent years progress has been made toward understanding the complex interrelationships between photoperiods, gonadal development and hypothalamic-pituitary secretions, and some of these findings are reviewed below.

It is recognised that the onset of the breeding season is brought about by the activation of the hypothalamic-pituitary axis by photic stimuli. During periods of anoestrus or testicular regression basal levels of gonadotrophins are extremely low and only infrequent low-amplitude pulses of LH secretion can be measured. The onset of reproductive activity is marked by a gradual increase in the frequency and amplitude of LH pulses and a slow but delayed rise in FSH levels. The increased gonadotrophin secretion consequently stimulates the gonads and the synthesis of testosterone and oestrogen. Careful analysis of the changing patterns of LH secretion which occur during the development of the testes from its regressed state to a fully active state have been studied in the ram (for details see [8]). At the time when the gonads are developing in response to increasing levels of gonadotrophins and the testosterone levels are rising in parallel, the amplitude of the LH pulses is high. However, when the testes are fully active and secreting high concentrations of testosterone, the amplitude of the pulsatile LH secretions is decreased while the duration is prolonged. It has been proposed that during the developing phase the synthesis of LH outweighs its release and thus the pituitary contains a relatively large pool of readily releasable LH. Conversely during the active phase the release of LH outweighs its synthesis thereby reducing the concentration of releasable LH and hence the amplitude of the LH pulse (see Figure 9.3). This theory is supported by the finding that the maximum

Figure 9.3: Photoperiodic Control of Gonadotrophin Secretion in the Soay Ram. (a) Increase in testicular diameter induced by an abrupt change in artificial photoperiods. (b) Changes in the frequency and amplitude of the LH episodic secretory patterns measured during the regressed, developing and active state of the testes. (c) Diagrams of the postulated changes in the releasable and reserve pools of pituitary LH. (d) The LH response to a bolus intravenous injection of 100ng Gn-RH. The amplitude of the LH release reflects the pituitary responsiveness and the duration of release reflects the size of the reserve pool. (Adapted from Lincoln, G. A. (1976) *J. Endocrinol.*, **69**: 213, with permission).

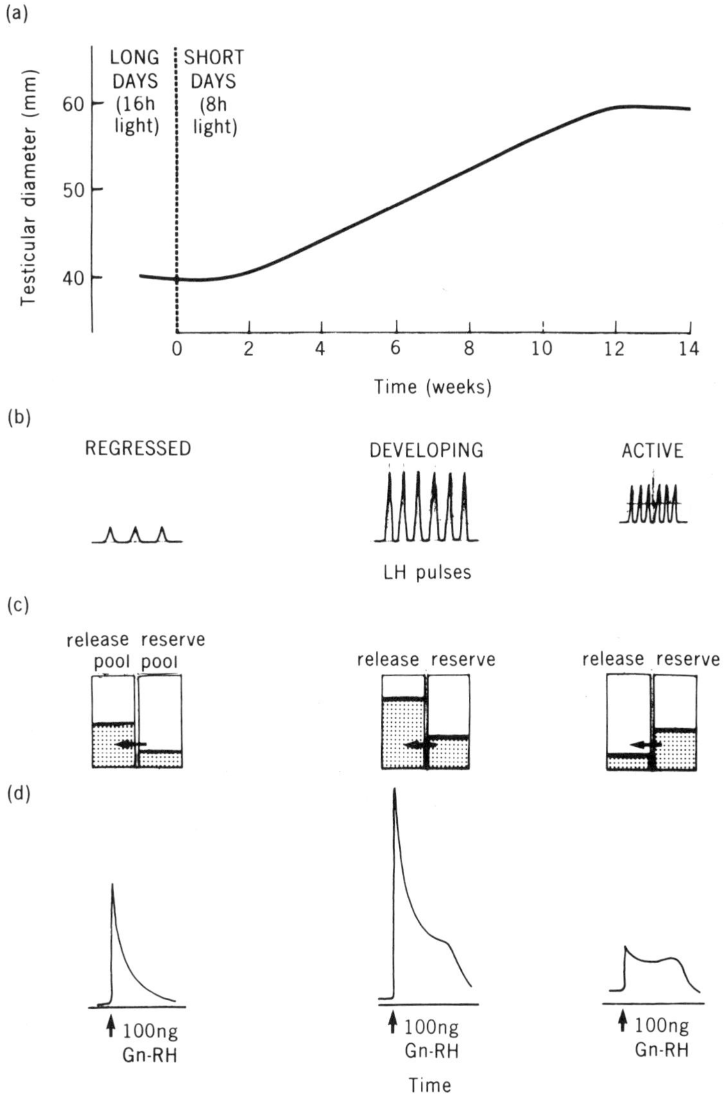

Figure 9.4: (a) Seasonal Change in Response to the Negative Feedback Action of Oestradiol on LH Secretion in Ewes. Ovariectomised (OVX) ewes show raised levels of LH throughout the year due to loss of steroid feedback but with oestradiol implants (which maintain a constant circulating concentration of the steroid throughout the year) the negative feedback effects are high during the anoestrous season and thus LH secretion falls to undetectable levels. (b) Histogram showing the breeding and anoestrous seasons in a group of intact ewes. (c) Hypothetical scheme for the control of seasonal breeding indicating the changes in the feedback effects of oestradiol on LH secretion. (Adapted from Karsch, F. J. *et al*. (1980) *J. Reprod. Fertil.*, **58**: 521, with permission.)

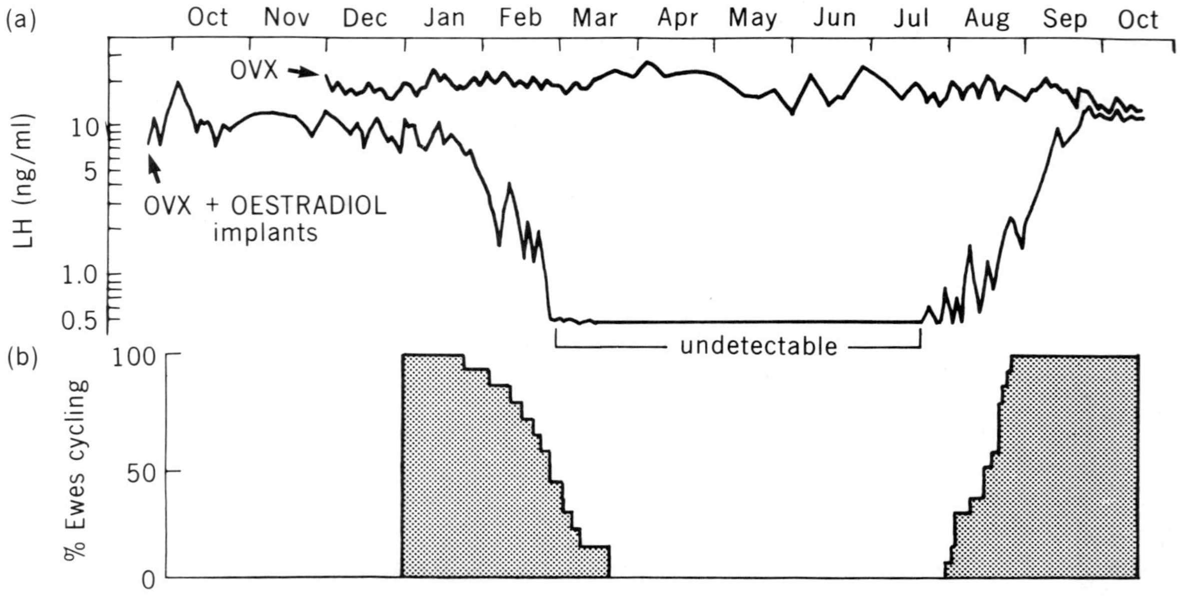

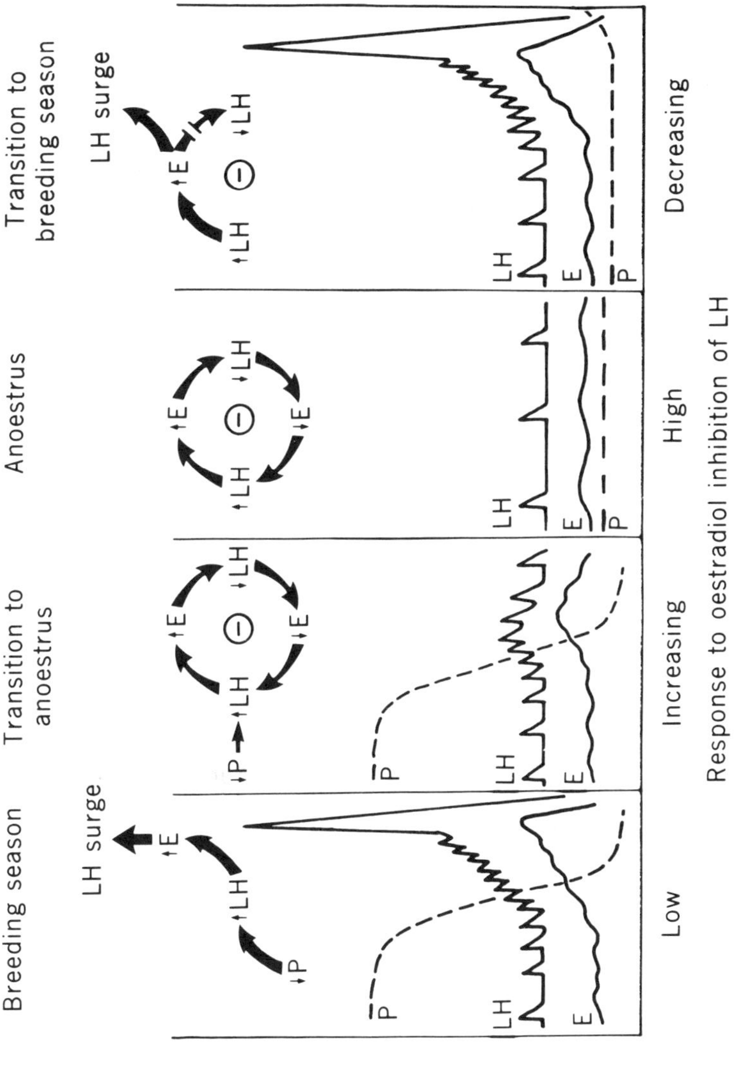
Breeding season
LH surge
Transition to anoestrus
Anoestrus
Transition to breeding season
LH surge
P
LH
E
Low
Increasing
High
Decreasing
Response to oestradiol inhibition of LH

(c)

pituitary content of the gonadotrophins is observed during the weeks prior to mating with a temporary depression occurring in the mid-mating season when reproductive activity is at a maximum. The pituitary content of gonadotrophins is minimal during the non-breeding season which probably reflects the lack of Gn-RH stimulation on the synthesis of LH and FSH. The responsiveness of the pituitary to Gn-RH also shows seasonal variations and parallels the changing levels of LH (see Figure 9.3(d)). This change may be attributed to the frequency of Gn-RH pulses reaching the pituitary coupled with the feedback effects of steroids on the gland. Interestingly, recent studies have shown that chronic 2 hourly pulsatile infusions of Gn-RH, but not continuous infusions, can faithfully reproduce the patterns of hormone secretion observed during the natural onset of the breeding season. Such studies strongly emphasise the importance of pulsatile Gn-RH secretion in this process and it is likely that changing photoperiods can activate neural mechanisms controlling episodic Gn-RH release, thereby stimulating the onset of seasonal reproductive functions.

A conceptual model for the control of seasonal breeding in the ewe (see Figure 9.4) is based on the idea that the neural centre controlling tonic LH secretion shows altered sensitivity to the negative feedback effects of oestrogen during different reproductive states (for details see [3]). With the decreasing photoperiod of the autumn and early winter, the tonic control system becomes less responsive to the inhibitory effects of oestrogen, allowing pulsatile LH secretion to increase. This then drives the increase in oestradiol necessary to trigger the preovulatory LH surge. Subsequent ovulatory cycles (16-17 days in length) during the breeding season are initiated by the withdrawal of progesterone (following luteolysis), leading to the preovulatory sequence as described above. The transition from oestrus to anoestrus, as a result of lengthening photoperiods, is brought about by an increased sensitivity to the negative feedback action of oestrogen which inhibits the follicular rise of LH secretion. During periods of anoestrous, there is an augmented response to the oestradiol inhibition of tonic LH secretion, ovulatory cycles are therefore repressed and pulsatile release of LH is low and infrequent. However, it would be incorrect to assume that the individual components of the hypothalamic-hypophyseal-ovarian axis are non-functional during periods of anoestrous. Indeed ovulation can be induced during these periods if the ovary is presented with an appropriate gonadotrophic stimulus and both positive and negative feedback effects of oestrogen can be demonstrated experimentally in anoestrous animals.

One important question arising from this model of seasonal breeding in the ewe is whether or not seasonal variations of reproductive activity rely only on changes in the sensitivity of steroid feedback mechansisms. In the absence of gonadal steroids (i.e. after gonadectomy) only the ewe and red deer fail to show seasonal changes in LH secretion whereas in other species which have been investigated, the circulating levels of LH show photoperiodic dependent fluctuations even in the absence of gonadal steroids. Conversely, when oestradiol concentrations are experimentally maintained at high levels in ovariectomised ewes, a remarkable seasonal change in LH secretion is observed. During anoestrus LH levels are undetectable but a 20-fold increase occurs during the breeding season, despite the fact that over this 12 month period oestrogen concentrations remained stable (see Figure 9.4). One may therefore argue that, either the oestradiol response in cells which modulate tonic LH secretion is dependent on the photoperiod, or that photoperiods may exert a direct independent 'drive' on tonic centres, which does not involve changes in the sensitivity of steroid feedback mechanisms. This question remains to be resolved. One point that should be raised is the number of parallels which have been drawn between the onset of the breeding season and puberty. Both processes may involve a resetting of the 'gonadostat', in that a change in the sensitivity of the feedback action of steroids may permit increased gonadotrophin secretion and hence stimulation of gonadal activity.

The evolutionary significance of seasonal breeders is apparent if one considers that the process of natural selection would favour those species whose offspring are born at a time of the year which favours the growth and development of the young. The mechanisms translating photoperiodic cues into changes of hormone secretions are undefined, although it seems likely that alterations of light entrained circadian rhythms may play some role. Destruction of the pineal or lesions of the suprachiasmatic nuclei can disrupt photoperiodic responses as well as circadian rhythms and it may well be that the pivot for seasonal reproductive cycles rests on subtle shifts in light entrained circadian rhythms. For further reading on the photoperiodic and neuroendocrine regulation of gonadal activity and seasonal breeding in mammals see [3, 7, 8, 13].

References

1. Boyar, R. M. (1978) Sleep-related Endocrine Rhythms, in S. Reichlin,

R. J. Baldessarini & J. B. Martin (eds) *The Hypothalamus* (Raven Press, New York), pp. 373-85.
2. Cardinali, D. P. (1979) Models in Neuroendocrinology: Neurohumoral Pathways to the Pineal Gland. *Trends in Neurosciences*, **2**: 250-3.
3. Karsch, F. J., Goodman, R. L. & Legan, S. J. (1980) Feedback Basis of Seasonal Breeding: Test of an Hypothesis. *Journal of Reproduction and Fertility*, **58**: 521-35.
4. Klein, D. C. (1978) The Pineal Gland: a Model of Neuroendocrine Regulation, in S. Reichlin, R. J. Baldessarini & J. B. Martin (eds) *The Hypothalamus* (Raven Press, New York), pp. 303-26.
5. Kreiger, D. T. (1979) *Endocrine Rhythms* (Raven Press, New York).
6. Krieger, D. T. & Hughes, J. C. (eds) (1980) Various articles on Hormones, Light and Biologic Rhythms, Section II in *Neuroendocrinology – a Hospital Practice Book* (Sinauer Assoc. Inc., Sunderland, MA).
7. Legan, S. J. & Karsch, F. J. (1979) Neuroendocrine Regulation of the Oestrous Cycle and Seasonal Breeding in the Ewe. *Biology of Reproduction*, **20**: 74-85.
8. Lincoln, G. A. & Short, R. V. (1980) Seasonal Breeding: Natures Contraceptive. *Recent Progress in Hormone Research*, **36**: 1-52.
9. Moore, R. Y. (1978) Central Neural Control of Circadian Rhythms, in W. F. Ganong & L. Martini (eds) *Frontiers in Neuroendocrinology*, Vol. 5, pp. 185-206.
10. Rusak, B. & Zucker, I. (1979) Neural Regulation of Circadian Rhythms. *Physiological Reviews*, **59**: 449-520.
11. Szafarczyk, A., Hery, M., Waplante, E., Ixart, G. Assenmacher, I. & Kordon, C. (1980) Temporal relationships between the Circadian Rhythmicity in Plasma levels of Pituitary Hormones and in Hypothalamic Concentrations of Releasing Factors. *Neuroendocrinology*, **30**: 364-76.
12. Turek, F. W. (1981) Are the Suprachiasmatic Nuclei the Location of the Biological Clock in Mammals? *Nature*, **893**: 289-90.
13. Turek, F. W. & Campbell, C. S. (1979) Photoperiodic Regulation of Neuroendocrine-Gonadal Activity. *Biology of Reproduction*, **20**: 32-50.
14. Zatz, M. (1980) The Pineal Gland: Shedding Light on the Internal Clock. *Trends in Pharmacological Sciences*, **1**: 230-3.

10

NEUROPEPTIDES: HORMONES OR TRANSMITTERS?

10:1 Introduction

At the beginning of Chapter 1 we alluded to the wealth of neuroendocrine research which has occurred in the last few years, and how recent advances have imposed a requirement for new definitions and the development of new concepts in our understanding of neural and endocrine systems. This has partly arisen as a result of the discovery of numerous biologically active peptides in both the central and peripheral nervous systems, as well as in the alimentary tract and other tissues. The peptide hormones first isolated from the hypothalamus were found to be distributed in many regions of the nervous system, whilst peptides previously classified as gut hormones were shown to be present in brain. Thus peptide hormones and other active peptides are located throughout the body. However, the nervous system is particularly rich in biologically active peptides. The majority of these peptides have been shown to have a discrete distribution in neural tissue and this has given rise to the general term 'neuropeptides'.

In this chapter the distribution of the neuropeptides, their characteristic features, the activities and potential roles for the neuropeptides will be considered. The effects of neuropeptides on behaviour will be discussed in the Chapter 11 together with the behavioural effects of the steroid hormones. We shall also consider in the subsequent chapter the stability and transport of biologically active peptides in blood and consider the question whether peptides produced in peripheral tissues (or injected parenterally) are taken up into the brain and can thereby regulate neural and neuroendocrine function.

Table 10.1: Main Distribution of Neuropeptides

Peptide	Size (No. of amino acids)	CNS	Pituitary and other tissues
(A) *Neurohypophyseal peptides*			
Vasopressin	9	Supraoptic and paraventricular nuclei, suprachiasmatic nuclei. Low levels in mid-brain parts of limbic system, medulla and CSF.	Posterior pituitary
Oxytocin	9	Supraoptic and paraventricular nuclei, thalamus, spinal cord.	Posterior pituitary
(B) *Hypophysiotrophic peptides*			
TRH	3	Ventromedial nucleus, preoptic area, paraventricular nucleus, nucleus accumbens, septal regions, thalamus, brain stem, ventral spinal cord.	Median eminence, retina, pancreas gut, pineal gland, placenta.
Gn-RH	10	Preoptic area, OVLT, suprachiasmatic nucleus septal nucleus, arcuate nucleus, olfactory	Median eminence, gonads.
Somatostatin	14	Suprachiasmatic nucleus, arcuate nucleus, preoptic area, olfactory, stria terminalis, nucleus accumbens, septum, thalamus, amygdala, cerebral cortex, central grey, dorsal spinal cord.	Median eminence, posterior pituitary, pancreatic islets, gut, stomach, sensory neurones, sciatic neurones, retina.
CRF	41	Hypothalamus, paraventricular nucleus	Median eminence, posterior pituitary
(C) *Pituitary peptides*			
(i) α-MSH	13	Preoptic area, arcuate nucleus, thalamus, striatum septum, brain stem, dorsal spinal cord.	Intermediate lobe, posterior pituitary? pineal gland
(ii) ACTH	39	Preoptic area, arcuate nucleus, septum, thalamus, hippocampus, amygdala, cerebral cortex, stria terminalis, periaqueductal grey.	Anterior pituitary, intermediate lobe, stomach, gut, pancreas, placenta.

Table 10.1 continued

Peptide	Size (No. of amino acids)	CNS	Pituitary and other tissues
(iii) β-endorphin	13	Olfactory, corpus callosum septum, preoptic area, thalamus, arcuate nucleus, interpenduncular nucleus habennula, spinal cord.	Anterior pituitary intermediate lobe, sympathetic ganglia, pineal gland.
(iv) β-LPH	91	Similar distribution to ACTH	as ACTH
(D) *Other neuropeptides*			
Met-enkephalin (Leu-enkephalin)	5	Globus pallidus, central grey, nucleus accumbens, amygdala, hippocampus, medial forebrain bundle, dorsal spinal cord.	Posterior pituitary; gut, adrenal medulla.
Substance P	11	Striatum, substantia nigra, amygdala, globus pallidus, preoptic area, dorsal spinal cord.	Gut, myenteric plexus, sensory neurones, sciatic and vagus nerves.
Neurotensin	13	Basal ganglia, preoptic area, nucleus accumbens, thalamus, amygdala, dorsal spinal cord.	Median eminence, anterior and posterior pituitary, gut (ileum).
Angiotensin II	8	Paraventricular nucleus, amygdala, CSF.	Blood, median eminence, anterior and posterior pituitary.
(E) *Peptides of Alimentary Tract*			
VIP	28	Stria terminalis, CSF, amygdala, cerebral cortex, hippocampus	Stomach, gut, vagus and sciatic neurones.
CCK	4, 8	Cerebral cortex, hippocampus, amygdala, stria terminalis, basal ganglia	Gut and pancreas
Insulin	51	Brain (?)	Pancreas

All neuropeptides listed above have been localised to various regions of the hypothalamus, including the medial basal hypothalamus. Where specific hypothalamic nuclei are listed refer to Figure 1.2. Other peptides which have been reported to be present in brain include the pituitary hormones, prolactin, growth hormone, gonadotrophins, and thyrotrophin, the gut hormones, gastrin, glucagon, and pancreatic polypeptide and peptides found in amphibian skin and other animal tissues including bradykinins, tachykinins and carnosine.

OVLT – organum vasculosum lamina terminalis

10:2 Localisation and Characteristic Features of Neuropeptides

Through the 1970s it became increasingly apparent that the number of chemical messengers in the central nervous system (CNS) far exceeded earlier predictions and many of the 'new' possible neurotransmitter substances were shown to be neuropeptides. Neuropeptides consist of single chains of amino acids, the number ranging from two to more than fifty (see Table 10.1) and are structurally identical or similar to peptides found in the peripheral nervous and endocrine systems. To date more than 20 neuropeptides have been described although the list continues to grow and predictions of the total number would certainly be premature at this stage of neuroendocrine and neuropeptide research. As a result of their isolation, chemical characterisation and synthesis, antibodies to the peptides were produced and consequently appropriate radioimmunoassay and immunocytochemical methods developed. These immunological techniques have enabled detailed analysis of peptide localisation in brain areas and demonstrated a subcellular distribution in nerve fibres, nerve terminals and occasionally neurone cell bodies. Although a detailed and complex distribution has emerged, the results obtained by these methods have to be interpreted with caution since they depend on the specificity and sensitivity of the antibodies in use. Information on the 'cross reactivity' of antibodies raised to a particular peptide may be obtained by prior testing with various pure synthetic peptide and peptide fragments. Peptide-producing tissues and biological fluids however, are known to contain both peptide precursor and processed or metabolised fragments of the natural peptide and, in addition, unrelated, but structurally similar peptides, which the particular antibody may recognise. To overcome the difficulties of peptide identity a series of antibodies may be used where each is known to recognise different parts of the peptide sequence. This approach has been particularly important in resolving the ACTH-LPH related peptides (see Chapter 5 Section 2) but total resolution of peptide identity in biological material will ultimately require more refined alternative techniques. Using well characterised antibodies, different laboratories have generally obtained similar pictures of neuropeptide distribution although quantitative discrepancies occur.

All the identified peptidergic hypothalamic hormones and some pituitary hormones have been found in brain as have most local hormones of the alimentary canal (Table 10.1) (see [16]). In addition very many neuropeptides have been previously (or in a few cases subsequently)

Table 10.2: Neuropeptides in Brain, Gut and Skin

Mammalian Brain and Gut	Amphibian Skin
Bradykinin-like	Bradykinins (8)
Substance P	Tachykinins (6)
Bombesin-like	Bombesins (8)
CCK/gastrin-like	Caerulein (3)
Enkephalins and Endorphins	Demorphin (2)
Angiotensins	Crinia-Angiotensin
Neurotensin	Xenopsin
CRF	Sauvagine
VIP	VIP(?)
TRH	TRH

isolated from amphibian skin tissue (Table 10.2) (see [5]), and some have been identified in various other vertebrates, a few invertebrates and even in plants. The significance of the common identity of neuropeptides in vertebrate brain, gut and skin will be discussed below (see Section 4), but it appears that these peptides may serve many roles in different tissues and throughout evolution similar molecules have been adapted to different functions at different times. A number of general features of the neuropeptides have been indicated by the immunocytochemical detection techniques. The peptides are present in discrete groups of nerve cells such that certain brain regions are richly endowed with neuropeptides (see Table 10.1) whereas others are relatively free of known peptides. It may be pertinent to note however, that whilst peptides are readily detected by immunocytochemistry in nerve terminals, their detection in nerve cell bodies has been more difficult. One reason for this is that lower concentrations of the peptide are present in the soma (cell body) compared with nerve endings where the peptide is stored and precursor forms of the peptides present in the neurone cell bodies may not be recognised by the specific antibody for the processed peptide. Because of this and other technical difficulties (see Chapter 2 Section 2(ii)), our knowledge of the distribution of peptidergic neurones may be rather incomplete. Nonetheless, it can be seen from Table 10.1 that the majority of the known neuropeptides are found in the medial basal hypothalamus and, generally, very high levels are found associated with nerve terminals in the stalk median eminence. Such a distribution suggests that many neuropeptides may have a neuroendocrine function and increasing evidence supports this idea (see Section 5). Several hypothalamic nuclei that contain neurones which project to the median eminence also have projections to other extrahypothalamic brain areas and it may be that hypothalamic neurones which contain neuropeptides are important for the integration of hypothalamic function with other central nervous system functions. For example neurone cell bodies containing the ACTH-like peptides – α-MSH, β-lipotrophin and β-endorphin occur only in the arcuate nucleus of the hypothalamus and the presence of these peptides in nerve terminals in other areas of the brain may be due to projections from these cells.

The hypothalamic paraventricular and supraoptic nuclei which synthesise the posterior pituitary peptides, vasopressin and oxytocin (see Chapter 3 Section 2) also contain other neuropeptides and just anterior to the hypothalamus, the preoptic area (POA) and the suprachiasmatic regions (important in the control of reproductive functions and circadian rhythms; see Chapter 9 Section 2) are richly endowed with peptidergic neurones. Regions of the basal ganglia and the limbic system are other brain areas which contain relatively high concentrations of neuropeptides and there is increasing evidence for many extrahypothalamic systems of peptidergic neurones. For further reading on neuropeptide distribution see [1, 2, 3, 9, 10].

Knowledge of the localisation of neuropeptides coupled with the evidence that peptides are released from neurones and affect neuronal activity or specifically alter behaviour (see Chapter 11 Sections 5–9) has created much discussion whether the neuropeptides may act either as neurotransmitters or as some form of specific neuromodulator of other known transmitters (see Section 3). Irrespective of their role in the CNS, an intriguing characteristic of neuropeptides is their apparent association with particular physiological functions (see Table 11.1). As mentioned previously (Chapter 2 Section 3(iii)) there is evidence that the hypothalamic neuro-secretory neurones have colateral axons which presumably release hypophysiotrophic peptides into other brain regions, as well as axon terminals on the hypophyseal portal vessels to regulate pituitary function. This means that a peptide from the same source may have both endocrine and neural activity. Although present evidence is by no means conclusive, we know for example, that the gonadotrophin releasing hormone, (Gn-RH) elicits sexual behaviour (see Chapter 11 Section 8) and regulates the release of gonadotrophin hormones, whereas thyrotrophin releasing hormone (TRH) stimulates thermogenic mechanisms and arousal in brain (see Chapter 6 Section 4 and Chapter 11 Section 7) in addition to its effects on TSH secretion. It is thus possible that these hypophysiotrophic neuropeptides function as either hormones or neurotransmitters, in different tissues, mediating a particular adaptive response of the animal.

Very little is known about the biochemical mechanisms of synthesis of neuropeptides, with the exception of the hypothalamic neurohypophyseal hormones (see Chapter 3 Section 3) but the latter are likely to be representative of the processes for neuropeptides in general. There is evidence that several neuropeptides are derived from larger precursor polypeptides. These have been isolated and chemically characterised in a few cases but the true relationships of isolated structurally similar peptides are far from clear. Peptide precursors are thought to be processed into smaller peptides by the action of intracellular peptidase enzymes prior to release but detailed analysis of such processes are particularly difficult since it is clear that similar peptide processing can occur after release as a result of the action of extracellular peptidases. These peptidase enzymes appear to be widespread in brain and other tissues where they are thought to provide a mechanism of rapid inactivation of peptide at target sites (see Section 3). Similar enzymes are active in blood resulting in very short half lives for the availability of neuropeptides and the possible significance of this is considered in the next chapter.

10:3 Neuropeptides as Neurotransmitters

The strongest candidates among the neuropeptides for neurotransmitter status are substance P, closely followed by the enkephalins. Of the hypothalamic hormones, perhaps the best evidence for neurotransmitter function rests with TRH, followed by Gn-RH but some evidence is available for somatostatin, neurotensin, vasopressin, cholecystokinin (CCK) (see [17]) and others. For details on neurotransmitter neuropeptides see [4, 12]. Such evidence is based on a number of criteria which may be used to attempt to establish neurotransmitter function. These may be summarised as follows:

(a) The substance is synthesised within the neurones in which it is localised and the appropriate enzymes, precursors and substances for synthesis must be present (see Chapter 2 Section 2(ii)).

(b) The substance is stored within the neurones where it is usually concentrated within synaptic storage vesicles or granules (see Chapter 2 Section 2(ii)).

(c) The substance is released from the nerve terminals following physiological stimulation by a calcium-dependent process common with that for most neurotransmitter and hormone release mechanisms (see Chapter 2 Section 3(v)).

(d) The substance must interact with the postsynaptic cell through a specific receptor mechanism (see Chapter 2 Section 5(ii)).

(e) A mechanism for the rapid termination of the action of the released substance must be present. This may take the form of an enzymic inactivation or a presynaptic uptake process (see Chapter 2 Section 5(iii)).

(f) The substance, when applied exogenously to the neurones, must mimic the actions of the true (endogenous) transmitter when this is released by electrical or physiological stimulation (see Chapter 2 Section 3 (iii)).

In general criteria (a), (b), (c) and (e) are supported for most neuropeptides listed in Table 10.1. It has been demonstrated that the release of neuropeptides from brain tissue *in vitro* usually depends on the presence of calcium ions. Various peptidase enzymes, which rapidly inactivate or degrade released neuropeptide, are present in brain, blood and other tissues and recent evidence indicates that many such enzymes act on common peptide linkages in various neuropeptides. With respect to criterion (d), although receptors (specific binding sites) for most neuropeptides have been described and partially characterised, the study of receptor mechanisms for neuropeptides in brain, in pituitary and other sites is in its infancy and awaits further development of appropriate pharmacological tools, particularly specific receptor agonist and antagonist drugs. The greatest progress in this respect has been made with the enkephalins, the endogenous opioid peptides, whose receptors have been studied for several years using the narcotic, morphine, the opiate antagonist, naloxone, and other related drugs. However, very few potent agonist or antagonist analogues of neuropeptides have yet been produced despite some early successes with LHRH (Gn-RH) analogues (see Chapter 11 Section 4 and Chapter 12 Section 3). In addition, true receptor mediation requires a demonstration that peptide binding is associated with appropriate changes in biological activity, and apart from the opioid peptides this has not been demonstrated for many neuropeptides.

Immunohistochemical studies have led to the identification of peptidergic pathways in the CNS and peripheral nervous system. Various substance P neuronal pathways have been described in the CNS and peripheral neurones (Table 10.1). For example, it is well established that a significant proportion of the 'pain-conducting' sensory neurones which project to the dorsal horn of the spinal cord release substance P as their transmitter (see [8]). In the dorsal spinal cord the release of substance P is inhibited pre-synaptically by interneurones which themselves release enkephalin and thus this is one site where the endogenous opiate peptides and administered morphine drugs may act as part of their analgesic action. A substance P pathway well described in the brain is in the basal ganglia extending from cell bodies in the caudate nucleus or globus pallidus to nerve terminals in the substantia nigra. This pathway is considered important in the control of movement and has been shown to be seriously depleted in the post-mortem brains of Huntington's chorea patients who exhibit characteristic symptoms of abnormal dance-like movements. Further substance P pathways of unknown function have been reported and for details of these and other peptidergic pathways the reader is referred to the reviews in [1, 2, 4, 9]. Descriptions of neural pathways in the brain for other neuropeptides, including the hypothalamic hormone peptides such as Gn-RH, are only just beginning to emerge, generally as a result of recent, more refined cytochemical methods. Finally, it should be pointed out and perhaps emphasised that for most putative neurotransmitter peptides at the present time, criterion (f) remains to be convincingly demonstrated.

Electrophysiological studies, however, have contributed to the evidence that peptides act as neurotransmitters or neuromodulators. Identified neurones in the hypothalamus, spinal cord and other brain regions have been shown to alter their patterns of firing in response to specific peptides applied microiontophoretically. The effect of the hypophysiotropic peptides and other neuropeptides on the excitability and behaviour of neurones in brain and spinal cord will be considered in the next chapter together with the specific effects shown by these peptides on overall behaviour (see Chapter 11 Section 5).

The utilisation of a peptide as a neurotransmitter differs quite profoundly from the utilisation of a monoamine (e.g. dopamine) or an amino acid (e.g. α-aminobutyric acid). The latter transmitter substances can be readily synthesised from an amino acid obtained in the diet, and enzymes in the cytoplasm of neurones enable rapid synthesis, via a short series of steps. Many neurotransmitters can therefore be synthesised in nerve terminals where they are stored in vesicles prior to their release. In addition, there are generally active mechanisms for the reuptake of the amine or amino acid neurotransmitter after its release, thereby reducing the requirement for new synthesis. In contrast, in neurones that produce neuropeptides the process is more complex, since peptide synthesis occurs on the ribosomes which are present only in the cell bodies (see Chapter 3 Section 3). Thus the site of release of a peptide neurotransmitter from an axon terminal is remote from the place where the peptide is made. It may follow then, that neurones which release peptides may be unable to act repeatedly on another cell in a short span of time in the manner of monoamine neurones, which make their transmitters locally and have fewer problems in replenishing stores. It is possible, however, that peptidergic neurones possess complex mechanisms of storage analogous to the posterior pituitary neurones (see Chapter 3 Section 5) which would serve to compensate for this, and furthermore peptide terminals may be able to replenish their peptide stores following reuptake of peptide fragments; a process which has been tenuously suggested from recent preliminary evidence.

Another way in which peptide neurotransmitters may differ from monoamine transmitters, for example, may lie in their type of action on the membrane receptors of the target cells. Until the discovery of neuropeptides it was usually emphasised that neurotransmitters alter the permeability of the membranes of the target cells to sodium, potassium or calcium ions. As a result, the potential across the membrane of neurones would change thereby making the cell more excitable (an excitatory transmitter) or less excitable (an inhibitory transmitter). However, neurophysiological studies have shown recently that certain neuropeptides do not always alter the voltage gradient across the cell but simply prevent the depolarising or hyperpolarising effects of other excitatory and inhibitory neurotransmitters on the membrane potential (see Chapter 11 Section 5).

Also, studies incorporating electron-microscopic examination have shown that various peptidergic neurones in brain, common with monoaminergic neurones (e.g. certain noradrenergic neurones which innervate the cerebral cortex), do not directly synapse with other cells. This suggests that such neurones release their product diffusely to activate various cells in the vicinity and such a process is consistent with a general neuromodulator role for the peptides. It is interesting to speculate on the complex interactions which may occur and furthermore on the significance of the co-existence and presumed simultaneous release of neuropeptides and monoamine neurotransmitters from nerve terminals which occurs in some regions (see Section 4).

10:4 Co-existence of Neuropeptides

A significant number of neuropeptides have been localised in neurones which synthesise and release a classical neurotransmitter, usually a catecholamine (see Table 10.3). This co-existence of peptides and transmitters in the nervous system is a relatively recent discovery resulting from sophisticated advances in cythochemical methodology (see Chapter 2 Section 2(ii)). Although functionally very little is known concerning the interaction of co-existing peptides and transmitters, it would appear that both may be released following physiological stimulation and indeed in the adrenal medulla, for example, the co-existing enkephalins and catecholamines are stored and released from identical storage granules. In the peripheral nervous system, vasoactive intestinal polypeptide (VIP) (see [6]) and acetycholine co-exist in secretomotor neurones which innervate sweat glands and here a functional significance has been proposed. Whilst released acetylcholine directly stimulates the gland, the concomitant release of VIP from different regions of the nerve-ending causes vasodilatation of the capillary blood vessels, thereby indirectly assisting sweat gland function. It is impossible to predict accurately the significance of co-existence in the CNS or indeed to be certain that concomitant release of peptide and amine transmitter occurs. Preliminary studies from neuropharmacological studies have indicated peptide modulation at amine transmitter receptor sites but very little can be concluded at the present time. The release of peptide and neurotransmitter may in fact prove to be a generalised feature of neural tissue, but even if this is not the case, the previously held concept, usually referred to as Dale's Principle, which suggests that *one* neurone produces and releases *one* neurotransmitter, may be in need of some revision, if both co-existing amine and peptide act as neurotransmitters (see Section 3).

The co-existence of peptides and amines in endocrine cells has been known for a number of years, particularly in the gastro-intestinal tract and nervous system. During the 1960s A. G. E. Pearse observed that some neurones and endocrine cells producing peptide hormones shared a set of cytochemical features and ultrastructural characteristics. He coined the term APUD (Amine Precursor Uptake and Decarboxylation) cells to describe these peptide-containing cells which also contained amines and/or processes for amine uptake and decarboxylation. The APUD concept postulated that both neural and endocrine cells were derived from neural ectoderm, and so cells of a similar derivation may

Table 10.3: Co-existence of Neuropeptides with 'classical' Neurotransmitters

Peptide	Classical transmitter	Location	Species
Enkephalin	Adrenaline Noradrenaline	Adrenal medulla	Various
Enkephalin	Dopamine	Carotid body	Various
Somatostatin	Noradrenaline	Adrenal medulla	Human
Somatostatin	Noradrenaline	Sympathetic ganglia	Guinea pig
Neurotensin	Noradrenaline	Sympathetic ganglia	Cat
CCK	Dopamine	Ventral tegmentum	Rat, human
Substance P	5-Hydroxytryptamine	Medulla/ventral spinal cord	Rat
TRH	5-Hydroxytryptamine	Medulla/ventral spinal cord	Rat
VIP	Acetylcholine	Autonomic ganglia/ sweat gland	Cat

become either hormone-secreting cells in the gastro-intestinal tract, or neurones, or neurosecretory cells, or, in amphibian species, epidermal and dermal cells. The remarkable homology of biologically active peptides common to mammalian brain and gut and amphibian skin (Table 10.2) is consistent with the APUD concept (see Chapter 1 Section 12 and [16]). For further reading on neuropeptide co-existence in neurones see [2, 10, 11].

10:5 Effect of Neuropeptides on the Hypothalamic-Pituitary Axis

It is not unreasonable to expect that neuropeptides present in the medial basal hypothalamus (MBH) may be involved in the regulation of pituitary function. We have heard elsewhere in the book how the hypothalamic releasing and release-inhibiting hormones are transported to the pituitary target cells in the hypophyseal portal circulation and thereby exert marked and specific effects on the production and secretion of the pituitary trophic hormones. In addition to the hypophysiotrophic hormones many neuropeptides are present in the median eminence and these include vasopressin, neurotensin, angiotensin, CCK and VIP. Thus it is not surprising that they have biological effects on the hypothalamic-pituitary axis as described in previous chapters.

Table 10.4: Effects of Some Neuropeptides on Anterior Pituitary Hormone Release

Peptide	Response					
	ACTH	TSH	GH	PRL	LH	FSH
TRH	0	+	0	+	0	0
Gn-RH (LHRH)	0	0	0	0	+	+
Somatostatin	0	–	–	0	0	0
Angiotensin II	0	0+	–	–	0+	?
CCK-8	0	–	–	+	0	–
Gastrin	0	–	+	–	0	–
VIP	0	0	+	+	0	+
Substance P	–	0	+	–	0	+
Neurotensin	?	0	+	+	0	0
Met-enkephalin	+	–	+	+	–	–
β-Endorphin	–	–	+	+	–	–

+ = stimulatory; – = inhibitory; 0 = no effect. Some peptides act directly on the anterior pituitary gland, while others act within the hypothalamus to alter the secretion of the appropriate releasing or release inhibiting hormones.

Some of the neuropeptides may indeed act as hypophysiotrophic hormones, being secreted from the median eminence into the hypophyseal portal vessels. Significant concentrations of β-endorphin, neurotensin, VIP and vasopressin have been measured in hypophyseal portal blood and these peptides can affect pituitary secretions (see Table 10.4). VIP, for example, directly stimulates prolactin secretion at the pituitary lactotroph. In contrast the opioid peptides remove the tonic dopamine inhibition of prolactin release (see Chapter 7 Section 4). Assuming these neuropeptides act as neurotransmitters or neuromodulators, their effects might operate at various levels in the complex structures controlling pituitary function. They may act dendritically or presynaptically on the releasing-factor neurones, or may modulate diffusely the basal or stimulated release of neurohormones from the neurosecretory cells. Further well-documented examples of such interactions in the median eminence are the inhibitory effects of opiates and opioid peptides on the induced release of somatostatin and Gn-RH (see Chapter 8 Section 6(iv)), and of VIP on somatostatin release (see Chapter 7 Section 3(iv)). In addition several peptides including β-endorphin, enkephalins, VIP, CCK 8 and glucagon have been shown to alter the turnover rates of hypothalamic dopamine and noradrenaline and thus peptide effects on pituitary secretions may be mediated by changes in neurotransmitter availability. There is also evidence that the release of vasopressin and oxytocin from the neurosecretory nerve terminals in the neurohypophysis may be modulated by neuropeptides and Met- and Leu-enkephalin, angiotensin, neurotensin and CCK are present in the gland (see Chapter 3 Section 10).

The effect of neuropeptides on pituitary hormone secretion has been examined by (i) intraventricular administration of the peptide to animals and the detection of changes in plasma pituitary hormone, (ii) examining the effect of neuropeptide on the release of hypophysiotrophic hormone from incubated hypothalamic tissue and (iii) examining the effect of a neuropeptide on the release of pituitary hormones from isolated pituitary glands or cells. Some of the neuroendocrine effects of neuropeptides on pituitary hormone secretion are summarised in Table 10.4, which can only provide an indication of the complex interrelationships between neuropeptides, neurohormones and pituitary hormones. For further reading see [14, 15].

10:6 Present Conclusions and Future Developments

At the present time the true identity, diversity and activity of neuropeptides in both neural and neuroendocrine systems is still far from clear. From what has already emerged, however, it is certain that further interesting revelations are just around the corner awaiting future research. The generation of neuropeptides from larger precursor molecules, their further processing prior to release, and their degradation by peptidases after release will require careful analysis and make some questions particularly difficult to answer without the development of new or improved methodology. A further dimension in this challenging and expanding area of research relates to the interaction of peptides at their target receptor sites. Recent experiments in peripheral tissues indicates that a number of peptide hormones after interaction with cell-surface receptors are 'internalised' and preliminary studies have also indicated internalisation of Gn-RH and prolactin hormone–receptor complexes (see Section 1:10). Thus yet a further previously held concept that peptide hormones interact with cell membrane receptors and do not enter the target cell requires some revision. It remains to be seen whether similar processes occur with peptide hormones and neuropeptides in general.

Further detailed studies of co-existence are likely to uncover more complex, multiple systems where more than one peptide is processed from a common precursor as is well documented for the ACTH-related peptides (see Figure 5.3). For example, it has been shown already that some 5-hydroxytryptamine neurones in the medulla which project to the ventral spinal cord contain substance P, others contain TRH but some appear to contain both peptides in addition to the indoeamine. It is possible that the neuropeptides, substance P and TRH, produced in identical cells are in some way related and may be derived from a common precursor which has yet to be discovered. Finally, what is the nature of the precise roles of neuropeptides in the nervous and endocrine systems? We know that some are hormones, some are very likely to be neuromodulators or neurotransmitters, and several appear to have separate neural and endocrine functions in different tissues. But what are the mechanisms of action in brain? Are they short-term neuromodulators or modulators of cellular metabolism? Are some more long-term regulators of cell development and organisation as described for the steroid hormones (see Section 1:10)? Perhaps research undertaken in the near future will provide the answers and, more than likely, reveal further possibilities as yet unrealised.

References

1. Bloom, F. E. (1981) Neuropeptides. *Scientific American*, **245**; 114-24.
2. Burgen, A., Kosterlitz, H. W. & Iversen, L. L. (1980) Neuroactive Peptides (A discussion) Series of articles in *Proceedings of the Royal Society, B*, **202**: 1-195.
3. Elde, R. & Hokfelt, T. (1979) Localization of Hypophysiotropic Peptides and other Biologically Active Peptides within the Brain. *Annual Review of Physiology*, **41**: 587-602.
4. Emson, P. C. (1979) Peptides as Neurotransmitter Candidates in the Mammalian CNS. *Progress in Neurobiology*, **13**: 61-116.
5. Erspamer, V. & Melchiorri, P. (1980) Active Polypeptides: from Amphibian Skin to Gastrointestinal Tract and Brain of Mammals. *Trends in Pharmacological Sciences*, **1**: 391-5.
6. Fahrenkrug, J. (1980) Vasoactive Intestinal Polypeptide. *Trends in Neurosciences*, **3**: 1-2.
7. Guillemin, R. (1978) Peptides in the Brain: the New Endocrinology of the Neuron. *Science*, **202**: 390-402.
8. Henry, J. L. (1980) Substance P and Pain: an Updating. *Trends in Neurosciences*, **3**: 95-7.
9. Hokfelt, T. *et al.* (1978) Aminergic and Peptidergic Pathways in the Nervous System with Special Reference to the Hypothalamus in S. Reichlin, R. J. Baldessarini & J. B. Martin (eds) *The Hypothalamus* (Raven Press, New York), pp. 69-135.

10. Hokfelt, T., Johansson, O., Ljungdatil, A., Lundberg, J. M. & Schultzberg, M. (1980) Peptidergic Neurones. *Nature*, **284**: 515-21.
11. Hokfelt, T., Lundberg, J. M., Schultzberg, M., Johnasson, O., Ljungdahl, A. & Rehfeld, J. (1980) Coexistence of Peptides and Putative Transmitters in Neurones in E. Costa & M. Trabucchi (eds) *Neural Peptides and Neuronal Communication* (Raven Press, New York).
12. Iversen, L. L. (1979) The Chemistry of the Brain. *Scientific American*, **241** (3): 118-28.
13. Jackson, I. M. D. & Vale, W. W. (1981) Extrapituitary Functions of Hypothalamic Hormones (Symposium) Series of articles in *Federation Proceedings*, **40**: 2543-69.
14. McCann, S. M. (1980) Control of Anterior Pituitary Hormone Release by Brain peptides. *Neuroendocrinology*, **31**: 355-63.
15. Meites, J. (1980) Relation of Endogenous Opioid Peptides to Secretion of Hormones (Symposium) Series of articles in *Federation Proceedings*, **39**: 2531-54.
16. Pearse, A. G. E. (1976) Peptides in Brain and Intestine. *Nature*, **262**: 92-4.
17. Rehfeld, J. R. (1980) Cholecystokinin. *Trends in Neurosciences*, **3**: 65-7.
18. Synder, S. H. & Innis, R. B. (1979) Peptide Neurotransmitters. *Annual Review of Biochemistry*, **48**: 755-78.

(See also General Book References 1, 3, 5 and 18.)

11

HORMONES AND BEHAVIOUR

11:1 Steroid Hormones – Introduction

In previous chapters the action of steroid hormones in the brain and the regional specificity of steroid receptors has been discussed in relation to feedback mechanisms of these hormones in regulating pituitary hormone secretion. However, another important aspect of central steroid action concerns their effects on behaviour and it is the purpose of this section to review very briefly two classes of steroid hormones, namely the glucocorticoids and gonadal steroids whose effects on behaviour are well documented.

Steroid hormones are lipophilic planar molecules ideally designed for crossing cell membranes. Thus these hormones not only readily enter cells and act intracellularly (see Chapter 8 Section 5) but unlike peptides are also very easily taken up by the brain from blood to enter central neurones. It is not surprising therefore that the brain is an important target organ for steroid hormones and that certain behaviours are steroid modulated.

11:2 Glucocorticoids and Learned Behaviours

It has long been recognised that mood and affect are influenced by adrenocorticoid hormones and, since many factors which stimulate ACTH release are psychological in nature, it became of interest to investigate the regional distribution of glucocorticoid receptors in the brain and to study the consequence of hormone receptor interactions on behaviour. The binding of radiolabelled [^{3}H]corticosterone in the rat brain is, perhaps surprisingly, highest in the hippocampus septum and amygdala with apparently a lower receptor density in the hypothalamus and pituitary. However, dexamethasone, (a synthetic glucocorticoid) has a different pattern of binding with higher uptake concentrations in pituitary compared with brain (see Chapter 5 Section 6) which may explain the more potent ACTH-suppressing effect of dexamethasone compared with corticosterone.

The high uptake of glucocorticoids in the hippocampus obviously led to the idea that these steroids might alter behaviour related to hippocampal function. The hippocampus is functionally associated with short-term memory processes and it was shown that corticosterone could retard the extinction of a learned task whereas both adrenalectomy and hippocampal lesions facilitate extinction. These findings suggest that corticosterone may be inhibiting hippocampal functions, although such an inhibition appears to be selective since other behavioural parameters (e.g. exploratory behaviour) can be altered by hippocampal lesions but are not affected by adrenal steroids. It is also interesting to note that dexamethasone, in doses known to suppress ACTH secretion, is unable to mimic the effects of corticosterone on behaviour and this differentiation is in agreement with the binding distribution of labelled corticosterone and dexamethasone. Thus both uptake studies and behavioural experiments have provided evidence for corticosteroid modulations of brain functions and behaviour. What remains to be solved is the action of the steroid at the cellular level. The behavioural effects of the glucocorticoid are manifest within an hour of administration and although this time interval is rather short it can be explained by a 'genomic effect' of the steroid (see Chapter 1 Section 11 and Chapter 8 Section 5) i.e. a DNA-dependent increase in mRNA synthesis. Cellular events which have been linked with corticosterone action are an inhibition of hippocampal unit electrical activity and an increase in 5-hydroxytryptamine formation. These and other short latency effects of steroids in brain are unlikely to be mediated by genomic expression and indeed some can be demonstrated in isolated nerve-ending preparations in the absence of a nucleus. In addition a variety of sensory events and REM (rapid eye movement) sleep have also been shown to be influenced by the central actions of glucocorticoids and thus these steroids may have relatively diverse actions on brain function. For further reading see [4, 9] and K. Fuxe, J.-A. Gustafsson & L. Wetteberg (eds) (1981) *Steroid Hormone*

Regulation of the Brain (Pergamon Press) and V. H. T. James & J. R. Pasqualini (eds) (1980) *Hormonal Steroids* (Pergamon Press).

11:3 Gonadal Steroids and Sexual Behaviour

The importance of the gonadal steroids – androgens, oestrogens and progestins – in the maintenance of reproductive behaviour is indisputable and it is clear that steroid endocrine mechanisms, which regulate fertility in both sexes, influence sexual behaviour in such a way that the chances of fertilisation are maximal. Sexual behaviour is a basic instinctual drive which can be affected by a number of factors including environmental stimulation (e.g. tactile, olfactory, light), gonadal hormones and peptides (see Chapter 11 Section 8). It may also be shaped by social conditions and the social history of the individual. Thus steroid hormones are really one facet in the expression of sexual behaviour, and their action on the brain must be integrated with other neural mechanisms influencing sexual activity. In the adult, gonadal steroids promote sexual behaviour and these effects may be termed 'activational'. The same hormones also have important actions on the sexual development, not only on peripheral target tissues but on the central nervous system, where they appear to regulate the development of specific neuronal circuits which may modulate anterior pituitary secretions and sexual behaviour. Their actions on the developing nervous system may be classified as 'organisational' effects of steroids. The endocrine control of sexual behaviour comprises a large area of neuroendocrine research and only a cursory glance of this field will be presented in this section. The interested reader is referred to [1, 2, 6, 9, 14, 15] for more in-depth information.

(i) *Activational Effects*

In lower mammals hormonal activation of sexual behaviour is clear. Female animals show well-defined periods of sexual receptivity which are appropriately synchronised with their ovarian cycles. Conversely, sexual behaviour in male animals does not show cyclicity although seasonal variations of sexual activity in both sexes may be observed, and are related to the level of gonadal hormone secretion (see Chapter 9 Section 3(iii)). In some primates clearly defined periods of sexual activity can be observed, but in many primates and humans there are less well-circumscribed periods of sexual receptivity, and, with regard to humans, it is doubtful whether hormones play a significant part in comparison with the powerful effects of social learning.

In most mammals removal of the gonads results in a gradual loss of sexual behaviour in both sexes although, in primates and man, it appears there are neural mechanisms supporting sexual behaviour which are less dependent on steroid levels and can certainly override endocrine sensitive systems. However, even in these species loss of the gonads may attenuate sexual behaviour. Replacement therapy with oestrogen in the female and testosterone in the male will restore sexual behaviour whereas progesterone in the female may either synergise with oestrogen and elicit receptivity or may inhibit the action of oestrogen. For example, in rodents if progesterone is administered 24-48 hours after a subthreshold dose of oestrogen, sexual behaviour is elicited whereas if the progesterone is administered around the same time period as the oestrogen 'concurrent inhibition' of the sexual behaviour occurs. These observations have been correlated with oestrogen effects on brain progesterone receptors (see below). However, in some species, e.g. sow and cat, progesterone does not synergise with oestrogen to facilitate sexual behaviour.

In several species, brain electrical stimulation or lesion (see Chapter 2 Section 3(ii)) and steroid implant experiments have elucidated the precise steroid-sensitive brain regions involved in eliciting sexual behaviour. For instance, in the rat the anterior hypothalamus and medial preoptic area appear to be the major hormone sensitive sites for the activating effects of testosterone on sexual behaviour whereas the corresponding oestrogen-sensitive site in the female is the ventromedial nucleus of the hypothalamus. Less precise information is available for 'higher' mammals although this may reflect less dependency on steroid activation for sexual behaviour and a greater complexity of neural networks involved in eliciting such behaviour. Autoradiographic studies (see Chapter 2 Section 4(v)) have shown that in all mammalian species so far investigated there are nerve cells in the medial preoptic area, the medial hypothalamus and limbic forebrain structures which concentrate sex hormone (see Chapter 8 Section 5). Thus brain areas involved in eliciting sexual behaviour have high concentrations of both oestrogen and testosterone receptors. Receptors for progestins are found in the preoptic area and the ventromedial nucleus and it has been shown that these are inducible by oestradiol. Thus it appears that oestrogen 'priming' induces progesterone receptors and this may account for the activation of sexual behaviour by progesterone in oestrogen primed rodents.

In male animals there has been considerable discussion as to whether testosterone (5α-dihydrotestosterone) is the active hormone which induces sexual behaviour or whether it is converted into oestradiol by enzymic aromatisation (see Figure 8.5) prior to exerting its effects. Such enzymes are found in hypothalamus and cerebral cortex but *not* pituitary. In several subprimate species oestrogen can elicit sexual behaviour in male animals and drugs which inhibit the aromatisation of testosterone to oestradiol block the testosterone activation of sexual behaviour. Such findings suggest that, in some species studied, testosterone acts on the neural mechanisms underlying sexual behaviour only after its conversion into oestrogen. In both males and females the effects of gonadal steroids on mating behaviour in gonadectomised animals can be potentiated by the gonadotrophin releasing hormone (Gn-RH) although this hypophysiotrophic peptide has no effect on sexual activity in the absence of steroids. Further details of the action of Gn-RH on sexual behaviour are discussed in Section 8.

Precisely how steroids can activate neural circuits mediating sexual behaviour is unknown. There is a wealth of literature on the effects of central neurotransmitters on sexual behaviour, particularly the catecholamine transmitter substances, but results are equivocal and show wide variation between the species studied. There is some evidence that oestrogens in the hypothalamus can be hydroxylated to form catecholoestrogens (see Figure 8.6) which may interact with catecholamines and in this way modulate neuroendocrine and behavioural functions. However, because the catecholoestrogens are relatively unstable substances their possible roles remain uncertain. Other work has also indicated that peripheral effects of steroids may be important for inducing sexual behaviour, such as steroid-induced sensitivity changes in cutaneous sensory receptors associated with sexual stimulation. Thus the gonadal steroids appear to have both central and peripheral effects in activating sexual behaviour although they are but one factor in controlling the complex neural networks which initiate this basic instinct.

(ii) *Organisational Effects*

Gonadal steroids have marked effects on the development of the somatic and central nervous system as well as the development of the gonads, the duct system and secondary sexual characteristics. Such effects are classified as organisational. In many lower mammals, e.g. rodents, the secretion of androgens by the testes during foetal and postnatal development is a prerequisite for both the differentiation of the male reproductive tract and for the differentiation of the central nervous mechanisms, which control the acyclic male-type pattern of gonadotrophin release and male copulatory behaviour (see [6] in Chapter 8). If females are experimentally exposed to androgens during foetal or early neonatal life they are infertile in adulthood, showing no cyclic pattern of hormone secretion and no female type sexual behaviour (lordosis) in response to oestrogen and progesterone. Certainly in rodents, androgens appear to have an important role in organising neural circuits which are involved in eliciting stereotyped sexual behaviour and there is some implication of the involvement of oestrogen derived from testosterone in this mechanism (see Chapter 8 Section 5 and Chapter 11 Section 3(i)). The evidence of brain differences between male and female is not confined to patterns of behaviour or hormone secretions since anatomical differences have also been observed in the stucture and neural connections of certain preoptic areas and anterior hypothalamic nuclei (see [1]).

Extrapolation of these findings with rodents to other mammals is not straightforward. In male deer, castrated at birth, (thus depriving the animal of testosterone during postnatal development) some evidence of female-type behaviours are observed although mounting behaviour is retained, presumably due to neuronal imprinting by testosterone in the foetal stage. In sheep, treatment of the ewe with high doses of testosterone during pregnancy results in the male-type aggressive behaviour and loss of positive feedback effects of oestrogen (see Chapter 8 Section 4) in the female offspring. Similar treatment in monkeys will have the first effect (aggressive behaviour) on female offspring but not the changes in oestrogen feedback. Evidence for the organisational effects of sex steroids in humans usually comes from observations of abnormal hormone secretions resulting from a genetic defect. These include inability of target tissues to respond to androgens resulting in testicular feminisation, overproduction of androgens due to impaired steroid synthesis in the adrenal glands and an inborn inability to convert testosterone into its more potent metabolite, dihydrotestosterone (see Chapter 8 Section 6). Generally lack of androgens result in feminisation of the reproductive tract whereas excessive androgens masculinise reproductive development. However, the effects of steroids on central mechanisms, as opposed to peripheral development, are poorly defined simply because human sexual behaviour is impossible to quantify (one can hardly measure mounting attempts as used in animal studies!) and secondly because social conditioning of sex-related behaviour is difficult to separate from the organisational effect of steroids. In most species the development and maintenance of sexual behaviour is closely linked with gonadal steroid secretions and their actions both on the central nervous system and on the peripheral reproductive organs. Primates and humans show less dependency on steroids for sexual behaviour and their phylogenetically newer mechanisms for controlling sexual activity may have social rather than reproductive significance. For further reading on hormones and human sexual behaviour see [2, 6]. Also see Ciba Foundation Symposium on *Sex Hormones and Behaviour* (1979) No 62 and [6]. K. Fuxe, J. A. Gustafsson, & L. Wetterberg (eds) (1981) *Steroid Hormone Regulation of the Brain* (Pergamon Press).

11:4 Peptide Hormones – Introduction

The specific effects of steroid hormones on various behaviours have been recognised for many years but more recent studies indicate that numerous peptides, including the hypophysiotrophic and pituitary hormones and other neuropeptides (see previous chapter) markedly influence behaviour in animals and man and also alter neuronal activity in the central nervous system (CNS). The mechanism and sites of action by which peptides might affect behaviour, however, are clearly complex, since quite different responses are obtained when the peptides are administered by different routes, i.e. peripherally (intraperitoneal, intradermal, subcutaneous or intravenous) or centrally (intracerebroventricular or by direct microinjection into neural tissue) or even orally or intra-nasally. As a consequence, this is a very active area of research and many questions remain unanswered, but the importance of peptides in regulating behaviour is evident from the observations that specific peptides or groups of peptides are clearly associated with the modulation of behaviour associated with particular physiological responses (Table 11.1). Some of these will be discussed in this section.

In the chapter on neuropeptides the widespread distribution and localisation of many peptides in the CNS was described and it is perhaps not unreasonable to suppose that the effects observed in animals when peptides are introduced directly into the brain or the ventricles simply reflects the activity normally resulting from similar endogenous peptide. However, many behavioural studies have involved peripheral injection of peptides, and one reason why the effects obtained in such experiments may differ from those following central administration is that most peptides in the blood circulation, unlike the steroid hormones, are not readily taken up into the brain. But having said this,

Table 11.1: Neuroactive Peptides and Behaviour

Peptide	Behaviour
ACTH/MSH 4-10)	Improved attention, memory
Vasopressin	Improved attention, memory
TRH	General arousal, modulates affective disorders
MIF	Decreases Parkinsonian tremor, modulates affective disorders
Gn-RH (LHRH)	Induces sexual behaviour, increases libido
Enkephalins	Analgesia, modifies learning
Endorphins	Analgesia, modifies learning and sexual arousal
Angiotensin II	Thirst drive
δ-sleep inducing peptide	Induces sleep

we do know that certain substances essential to brain are able to cross what is sometimes, perhaps erroneously, referred to as the 'blood–brain barrier'. For further information see [4, 7]. It is pertinent to consider the various ways in which endogenous peptides in blood, or peptides injected peripherally, may have effects on the CNS. The peptide may act at a peripheral target site to release a mediator, e.g. a steroid, which readily enters and acts on the brain. Alternatively, the peptides may act at central sites, without entering the brain. For example, peptides may possibly have effects in specialised vascular regions called the circumventricular organs which are in close apposition to the CSF, since there is evidence that peptides and peptide receptors are present in these regions (see Chapter 11 Section 9). The highly vascular median eminence is, in fact, one such region, where material including peptides may be transferred between CSF and blood by specialised cells called the tanycytes (see Chapter 1 Section 8). Others include the organum vasculosum of the lamina terminalis (better known as the OVLT) which is rostral to the median eminence, the subfornical organ dorsal to the median eminence and the subcommissural organ caudal to the median eminence. See [15] for further information on the circumventricular organs.

It is possible, of course, that small quantities of some peptides in blood may enter the brain and this peptide may be sufficient to bring about the appropriate response since we know that very low concentrations of peptides administered centrally can cause marked behavioural changes in some instances. Finally, smaller peptide fragments or metabolites formed in blood may more readily cross into the brain. It is known that blood is richly endowed with peptidase enzymes and that small peptides in blood are rapidly processed into fragments or their constituent amino acids, such that the 'half life' (time required to reduce the concentration by half) for peptides is only a few minutes. Furthermore, active processes for the uptake of amino acids into brain exist in order to supply neurones and other cells with these essential substrates, and it may be that certain peptide fragments may use similar systems.

Because of the rapid degradation of neuropeptides by blood and other body tissues (especially in brain and liver) there is a constant vigilance by chemists to produce potent, active analogues which are more peptidase-resistant. Such analogues are invaluable for both endocrine and behavioural research studies as well as for potential future use as therapeutic drugs. Analogues are produced by substituting, deleting or adding appropriate amino acids or other chemical groups and where, as is often the case, the natural peptide has separate sites of endocrine and neural activities, the attempt is to obtain analogues with preferential potency for one activity only. Progress in this respect is proving slow since the 'receptors' or peptide recognition sites at the different target tissues are often apparently very similar. (For diagnostic and therapeutic value of hypophysiotropic peptide analogues see Chapter 12 Section 3.)

11:5 Peptides and the Activity of Central Neurones – Electrophysiological Studies

Before considering the effects of peptides on behaviour, it is useful to consider the evidence that many peptides affect the activity of brain and spinal cord neurones. This has been shown by electrophysiological studies using either *in vivo* micro-iontophoretic techniques, or *in vitro* methods which allow recording of electrical activity of perifused tissue (see Chapter 2 Section 3(iii)). By these techniques, minute quantities (femtomoles– 10^{-15}M) of peptide can be applied directly to the surface of single neurones. The hypophysiotrophic hormone peptides, thyrotrophin releasing hormone (TRH), gonadotrophin releasing hormone (Gn-RH) and somatostatin each exhibit both excitatory and inhibitory effects on identified hypothalamic and extra-hypothalamic neurones, and some of these activities have been consolidated using synthetic stable analogues of these peptides.

For example, TRH-sensitive neurones in rat brains, which are usually inhibited or show depressed excitability, are found in the medial basal hypothalamus, cerebral cortex, cerebellum and cuneate nucleus, whilst in the motorneurones of the ventral spinal cord TRH has a stimulatory effect. The central actions are consistent with behavioural parameters which indicate that TRH in brain has mostly excitatory effects which are likely to be mediated by disinhibition, since the peptide also antagonises the effects of inhibitory or hypnotic drugs (see Chapter 11 Section 7). Gn-RH responsive neurones, which are either stimulated or inhibited, are widely distributed in the CNS particularly throughout the hypothalamus and pre-optic area, but also in the septum, mid-brain central grey areas and cerebral cortex. Neurones in the pre-optic area and in the arcuate nucleus of the hypothalamus are mostly excited by Gn-RH and these observations complement our understanding of Gn-RH facilitated sexual behaviour in animals and humans (see Section 5). The Gn-RH containing neurones in these hypothalamic regions terminate in the median eminence and also have collaterals to other hypothalamic and extra-hypothalamic areas possibly to initiate or modulate various behaviours. Somatostatin-sensitive neurones are largely depressant and are found in the ventro-medial nucleus of the hypothalamus in particular but also in the cerebral cortex, hippocampus, cerebrellum and dorsal spinal cord. The behavioural significance of these neural effects of somatostatin is difficult to assess since the peptide is widely distributed throughout the CNS. Other neuropeptides also modify firing of central neurones. Among those which have been studied are the neurohypophyseal hormones, vasopressin and oxytocin, substance P, the enkephalins and endorphins. Substance P increases the neuronal excitability of selected neurones in spinal cord, brain stem, substantia nigra, cuneate nucleus, and cerebral cortex in particular. Characteristically the responses to substance P are of long duration and slow in onset and electrophysiological studies on the spinal cord and brain stem have added to the evidence that substance P is associated with pain sensory pathways (see Chapter 10 Section 3). Similarly the enkephalins and opiate drugs, which have been extensively studied with iontophoretic techniques, demonstrate a decrease in neuronal excitability in regions where they are considered to have analgesic activity; for example, in the substantia gelatinosa of the dorsal spinal cord where opioid peptides and drugs inhibit the release of substance P. Enkephalins, however, increase the excitability of neurones in the hippocampus and mid-brain regions and it is considered that, in this case, endogenous opioid peptide may activate descending systems which further modulate ascending sensory pain pathways. For further reading on the effects of peptides on the activity of central neurones see [10] and P. Collu, A. Barbeau, J. Ducharme & J. G. Rochefort (eds) (1979) *Central Nervous System Effects of Hypothalamic Hormones and Other Peptides* (Raven Press, New York), and P. D. Hrdina & R. L. Singhal (eds) (1981) *Neuroendocrine Regulation and Altered Behaviour* (Croom Helm, London).

11:6 Effects of Peptides on Memory and Learned Behaviours

A large body of evidence suggests that peptides which are derived from, or share a structural homology with corticotrophin (ACTH) and melanocyte stimulating hormone (MSH) (see Figure 5.3) increase attention or perceptual behaviour by central mechanisms which are not mediated by their peripheral endocrine activity. Furthermore, preliminary evidence indicates that the related endorphin peptides may have either similar or opposite effects but interpretation of the behavioural effects of the endorphins is not at all clear. In the early 1950s, following studies on the physiological regulation of ACTH secretion in response to various forms of stress, it was considered that fear and anxiety responses associated with stress were produced by ACTH. However, studies initiated in the mid 1960s by de Wied and others showed that ACTH, administered peripherally to rats, delayed the extinction of conditioned avoidance behaviour and this observation was interpreted as an effect of ACTH on enhancing short-term memory rather than stimulating fear. This view was further substantiated by experiments using less stressful, passive avoidance tests and carbon dioxide inhalation to induce amnesia. The work suggested that peptide fragments of ACTH (particularly $ACTH_{4-10}$) improves motivation, learning and short-term memory in animals and man. The same research workers have also reported that the neurohypophyseal hormone, vasopressin, similarly facilitates while oxytocin attenuates memory processes. In fact, vasopressin was the first peptide to be implicated in memory processes and has been recently marketed in certain countries as an anti-amnesic drug. For further reading see [4, 13].

More recent animal studies have implicated that posterior thalamic regions are important in the modulation of memory, and after testing many synthetic peptide fragments it appears that particular peptide sequences are more effective in attentional or acquisitional learning mechanisms, whilst others are more effective in improving consolidation of memory. It is assumed that these peptides are derived either from released pituitary hormones or from alternative sources in the hypothalamus and other brain areas, where ACTH-like and vasopressin-like immunoreactive material is produced independently of pituitary sources. There is evidence for similar behavioural effects of such peptides in humans based on a wide variety of behavioural and psychomotor tests. Such tests may be used to estimate consolidation of learning and acquisition, or memory retrieval processes, but the results of these tests are by no means conclusive or easy to interpret. In fact, there is an emergent body of opinion that the interpretation of the animal studies in terms of modulation of memory processes may be premature and the behavioural effects may simply reflect activation of specific arousal mechanisms (see Section 7). However, the potential value of using these peptides in the treatment of amnesia, particularly in the aged, is obvious and, although preliminary clinical trials have shown some beneficial effects, it will be necessary to obtain and vigorously test the behavioural efficacy of synthetic peptide analogues which are devoid of peripheral and endocrine side effects, before an anti-amnesic drug can be developed (see Section 11:4).

11:7 Effects of Peptides on Arousal and Sleep Mechanisms

Other neuropeptides, which do not directly alter attentional or motivational aspects of learning and memory processes, do act as CNS stimulants and generally enhance arousal behaviour, whilst others have the opposite effect. The hypophysiotrophic peptide TRH is widely distributed throughout a number of CNS regions (see Chapter 10 Section 2) and central or peripheral administration of TRH, or one of several biologically stable analogues of TRH (see Figure 6.3), to animals generally increases behavioural activity by mechanisms independent of its pituitary action (see [5, 12] in Chapter 6). For example, TRH potentiates the behavioural activation induced by monoamine oxidase inhibitor drugs (e.g. pargyline) with monoamine transmitter precursors (such as L-DOPA) in either hypophysectomised or thyroidectomised animals. TRH also antagonises sleep and hypothermia caused by alcohol or barbiturates, such as sodium pentobarbital, and opposes the action of several other depressant drugs. Furthermore, central administration of TRH temporarily increases locomotor activity and stimulates the excitability of motor neurones in ventral spinal cord. Consistent with an effect on arousal mechanisms, TRH administration also suppresses feeding and drinking behaviour, affects gut motility and gastric acid secretion and centrally activates a number of autonomic sympathetic mechanisms, including increase in heart rate and respiration and secretion of catecholamines from the adrenal medulla. In clinical trials TRH has been shown to have beneficial effects in depressed patients in some instances but in many others it has been ineffective in alleviating depressive symptoms. It has also shown variable relief in the treatment of schizophrenia, hyperactivity and alcohol withdrawal although to date there is no general acceptance of such a therapeutic value of TRH (see Chapter 12 Section 3).

In brain, TRH is metabolised to a diketopiperazine dipeptide (cyclo-His-Pro) (see Figure 6.3) and this metabolite has been shown to antagonise a number of central actions of TRH, as well as TRH-induced prolactin secretion from the pituitary. Recent evidence also indicates that a further neuropeptide, neurotensin, which is similarly distributed throughout brain, has reciprocal or opposing actions to TRH (see [11] for details). A further peptide, somatostatin, is largely inhibitory in brain and may act as a CNS depressant, while yet another recently isolated from rabbits acts as a hypnotic and induces sleep (delta sleep inducing peptides – see [8]). Much further research is necessary to consolidate the potential roles and mechanisms of action of these peptides which appear to modulate arousal or sleep mechanisms.

11:8 Peptides and Sexual Behaviour

The hypothalamic gonadotrophin-releasing hormone (Gn-RH), which induces the secretion of luteinising hormone (LH) and follicle stimulating hormone (FSH) from the anterior pituitary, might be considered as a likely mediator of sexual behaviour. Animal behavioural studies seem to confirm such a supposition since peripherally injected Gn-RH potentiates mating behaviour (lordosis) in oestrogen-primed ovariectomised rats which have been adrenalectomised or hypophysectomised, indicating a central action independent of gonads, adrenals, or pituitary gland. Similar effects have been observed in the mating behaviour of male rats as well as an acceleration of ejaculation. The Gn-RH-induced copulatory behaviour is dependent on the presence of testosterone in the male and oestrogen in the female (see Section 3). Similar studies have indicated that TRH inhibits mating behaviour and it has been shown that both the Gn-RH and TRH-sensitive sites for these behavioural effects are confined to the medial pre-optic area and the arcuate/ventromedial nuclei in the hypothalamus and the central grey in the mid-brain. In female rats Gn-RH is particularly active in the mid-brain central grey, an area of the brain from which descending axons originate and which are responsible for eliciting the lordosis response. The suggestion has been made that the ventromedial nucleus exerts an oestrogen-dependent tonic bias on the mesencephalic central grey and that Gn-RH neurones, also present in the ventromedial nucleus, may project to the same area. Thus in female rodents Gn-RH may act to synchronise the timing of ovulation with sexual receptivity thereby maximising the chances of fertilisation.

The observation that Gn-RH facilitates mating behaviour in rats has naturally generated considerable interest in the possibility that Gn-RH may be used to treat human disorders leading to sexual dysfunction and impotence. Interestingly, some clinical experiments and treatments of normal and impotent patients (particularly males) have reported beneficial changes in libido, whereas others have reported none. The potential clinical application of Gn-RH is discussed in Chapter 12 Section 3.

Preliminary reports have also suggested that β-endorphin may stimulate sexual arousal while other studies suggest an inhibition of sexual behaviour; but all these results remain to be substantiated. Melanocyte stimulating hormone (MSH) is produced and secreted from the pituitary intermediate lobe in rats and other mammals but its role in higher vertebrates is unknown. One suggestion for a role in sexual behaviour has emerged from recent studies. MSH has been shown to stimulate mating behaviour in rats and the mechanism suggested is the stimulation of pheromone secretion from the female rat preputial glands (see Chapter 4 Section 2(ii)).

11:9 Peptides and Other Behaviours

A number of other peptides have been variously reported to induce or modulate other forms of behaviour (see Table 11.1) and no doubt other examples will emerge in the future. At least one of these deserves further mention, namely angiotensin II, which perhaps demonstrates the most specific behavioural response when administered centrally to animals in minute amounts – namely a thirst drive causing immediate inducement to drink. This behavioural response has been shown in a number of species and possibly represents a further example of a common mediator regulating neural and endocrine components of an adaptive response, in this case to maintain body water balance (see [3, 15]). Angiotensin II is formed in blood by a series of processes initiated by renin release from the kidneys which is stimulated by a fall in blood pressure or sodium ions or by sympathetic stimulation. Angiotensin II has two peripheral actions, it stimulates aldosterone release from the adrenal cortex, which in turn causes sodium retention in the kidney, and is a potent vasoconstrictor hormone which assists in the restoration of blood pressure. Although a renin–angiotensin system is present in the brain, it has been considered that blood angiotensin may induce the thirst drive in brain by acting at the circumventricular regions where blood and CSF are in close proximity (see Section 4), since receptors for angiotensin II are present in one of these regions, the subfornical organ. Angiotensin II also has a further central action associated with water conservation in that it stimulates vasopressin secretion (see Chapter 3 Section 9(i)) and also the adrenal axis via CRF and corticotrophin release (see Chapter 5 Section 7). Hence angiotensin II may help to restore normal fluid balance and osmolality not only by a stimulatory action on the secretion of aldosterone but also on the thirst drive mechanism.

Whilst peptides have been shown to have specific behavioural effects in normal healthy individuals and animals, a number of examples have emerged in which certain peptides have been found useful in the diagnosis or treatment of particular diseases (see Chapter 12 Sections 1 and 3). One example of the latter is the use of the MSH release-inhibitory factor (MIF – see Chapter 4 Section 3) to reduce tremor and rigidity symptoms of Parkinson's disease. This is a relatively common neurological motor disorder associated with a depletion in dopaminergic neurones in the basal ganglia, although this is not considered the primary lesion. Most Parkinson patients are effectively treated with L-DOPA (a precursor of dopamine which readily enters the brain from blood) and early animal studies showed that MIF potentiated L-DOPA actions. Other peptides such as TRH, also show similar L-DOPA potentiation (see Section 11:7) but have proved relatively ineffective in Parkinson's disease.

References

1. Arai, Y (1981) Synaptic Correlates of Sexual Differentiation. *Trends in Neurosciences*, 4: 291-3.

2. Bancroft, J. (1981) Hormones and Human Sexual Behaviour. *British Medical Bulletin*, **37**: 153-8.
3. Buckley, J. P. (1981) Central Nervous System Effects of Angiotensin II. *Trends in Pharmacological Sciences*, **2**: 161-2.
4. De Kloet, R. & De Wied, D. (1980) The Brain as a Target Tissue for Hormones of Pituitary Origin: Behavioural and Biochemical Studies, in L. Martini & W. F. Ganong (eds) *Frontiers in Neuroendocrinology*. Vol. 6. (Raven Press, New York), pp. 157-201.
5. Dorner, G. (1980) Sexual Differentiation of the Brain. *Vitamins and Hormones*, **38**: 325-83.
6. Hutchinson, J. B. (ed.) (1979) *Biological Determinants of Sexual Behaviour*. (John Wiley & Son, Chichester).
7. Kastin, A. J., Olson, R. P., Schally, A. V. & Coy, D. H. (1979) CNS Effects of Peripherally Administered Brain Peptides. *Life Sciences*, **25**: 401-14.
8. Kastin, A. J., Olson, G. A., Schally, A. V. & Coy, D. W. (1980) DSIP – More than a Sleep Peptide. *Trends in Neurosciences*, **3**: 163-5.
9. McEwen, B. S. (1981) The Brain as a Target Organ of Endocrine Hormones, in D. T. Krieger & J. C. Hughes (eds) *Neuroendocrinology – a Hospital Practice Book* (Sinauer Assoc. Inc., Sunderland, MA), pp. 35-42.
10. Moss, R. L. (1979) Actions of Hypothalamic Hypophysiotropic Hormones on the Brain. *Annual Review of Physiology*, **41**: 617-31.
11. Nemeroff, C. B., Luttinger, D. & Prange, A. (1980) Neurotensin: Central Nervous System Effects of a Neuropeptide. *Trends in Neurosciences*, **3**: 212-5.
12. Ramsay, D. J. & Ganong, W. F. (1981) CNS Regulation of Salt and Water Intake in D. T. Krieger & J. C. Hughes (eds) *Neuroendocrinology – a Hospital Practice Book* (Sinauer Assoc. Inc., Sunderland, MA), pp. 123-9.
13. Rigter, H. & Crabbe, J. C. (1979) Modulation of Memory by Pituitary Hormones and Related Peptides. *Vitamins and Hormones*, **37**: 154-243.
14. Toran-Allerand, C. D. (1981) Gonadal Steroids and Brain Development. *Trends in Neurosciences*, **4**: 118-21.
15. Weindl, A. (1973) Neuroendocrine Aspects of the Circumventricular Organs in W. Gangong & L. Martini (eds) *Frontiers in Neuroendocrinology*, pp. 1-37.
16. Wilson, J. D. (1979) Sexual Differentiation. *Annual Review of Physiology*, **40**: 279-306.

(See also General Book References 4, 6 and 10; Ciba Foundation (1979) *Symposium on Sex Hormones and Behaviour*, no. 62. (Exerpta Medica, Amsterdam).)

12

CLINICAL ASPECTS OF NEUROENDOCRINOLOGY

The rapid development in the field of neuroendocrinology over the past decade has led to a newer understanding of human hypothalamic-pituitary dysfunction and has brought about the introduction of new diagnostic and therapeutic methods. The purpose of this chapter is to review briefly (a) the main types and possible causes of commonly presenting disorders, (b) pituitary function tests used in clinical practice, (c) the clinical use of hypothalamic peptides and related drugs which can modify the secretion of pituitary hormones and (d) to outline possible future developments in clinical neuroendocrinology, based on the current biochemical and chemical knowledge of the hypothalamic-pituitary axis.

12:1 Hypothalamic-Pituitary Dysfunction

Disorders of pituitary function are not as common as other endocrine diseases such as diabetes mellitus, and mostly result from the development of a tumour (e.g. an adenoma – see [8]). Pituitary dysfunction can also result from disorders of the hypothalamus, again via a tumour (e.g. a craniopharyngioma) but other causes include trauma after injury, vascular disorders, such as a pituitary infarct (pituitary apoplexy), subarachnoid haemorrhage, or very rarely, inflammatory problems resulting from infectious diseases such as encephalitis or meningitis. In addition there are rare cases of congenital disorders (e.g. diabetes insipidus or Gn-RH deficiency, see Section 2), a deficiency known as Sheenan's syndrome which occurs in mothers with post-partum haemorrhage and necrosis of the anterior pituitary.

Finally, there are iatrogenic causes resulting from surgery, drugs or irradiation treatment. The hypothalamic-pituitary disorders which concern the neuroendocrinologist manifest themselves as either hormone deficiency or alternatively as hormone overproduction and oversecretion.

(i) *Hypopituitary Function*

A single adenohypophyseal hormone deficiency is rare, but conditions of inappropriate growth hormone or gonadotrophin, or deficiency in both these hormones, occur most frequently and can be treated, particularly in the young, by appropriate hormone-replacement therapy. Gonadotrophin or steroid hormone replacement in children is generally only relevant where there is a delay in puberty and deficiency is confirmed, but in the adult, women in menopause may require oestrogen replacement therapy. Treatment with growth hormone has been limited by the availability of purified hormone, but the advent of the bacterial production of human growth hormone, and other large protein hormones using genetic engineering techniques will soon obviate such difficulties. Attempted treatment with somatomedins (see Chapter 7 Section 1(iv)) has not proved very successful, except in cases of low somatomedin and high growth hormone, but more information is required about these growth factors.

Although in congenital disorders it is more common to find a deficiency of one pituitary hormone in acquired disorders (e.g. due to a tumour), it is more common to find that the secretion of several hormones (hypopituitarism) or all anterior pituitary hormones (panhypopituitarism) are deficient. In clinical practice it is usually necessary to test for normal residual function of various hormones where deficiency in one is suspected (see Section 12:2). The frequency of hormone deficiency in descending order is usually gonadotrophins, growth hormone, TSH and ACTH whilst prolactin is very rarely lost. Hypopituitarism can lead to atrophy of the gonads, thyroid and adrenal glands, and hyposecretion of the respective hormones due to reduction in blood levels of these hormones. For further clinical and other details the interested reader is referred to [1, 4] and D. T. Krieger & J. C. Hughes (eds) (1980) *Neuroendocrinology – a Hospital Practice Book* (Sinauer Assoc. Inc., Sunderland, MA) and L. Martini & G. M. Besser (1977) *Clinical Neuroendocrinology* (Academic Press, New York).

Since a tumour is the most likely cause of a disorder, an initial investigation of the patient in order to confirm the presence of a tumour may involve radiography of the skull in the region surrounding the pituitary. In this way it is possible to search for evidence of displacement resulting from abnormal growth. Secondly, an examination of the patient's visual function is frequently undertaken since this may be disturbed because of the near vicinity of the optic chiasma. If a tumour is present and is sufficiently large it may cause varying degrees of hypopituitarism due to physical compression or disruption of normal function. A pituitary adenoma may be endocrinologically inactive or it may produce one or more hormones thus causing hypersecretion of particular hormones (see below). In addition to pituitary tumours secreting excess 'normal hormone' they may also secrete abnormal or 'foreign hormone' which may disrupt normal hormone action. A further possible endocrine complication may result from tumours which develop elsewhere in the body and secrete active pituitary hormones (ectopic secreting tumours) e.g. the oat-cell carcinoma of the lung often secretes ACTH-related peptides.

When hypopituitarism is suspected, irrespective of the cause, the primary endocrine considerations for the clinician are that the secretion of thyroid hormone and adrenal corticosteroid are normal, since deficiency of these hormones can lead to critical (even fatal) complications (Section 12:2). Because of normal circadian variations in cortisol, blood serum levels of this hormone are usually determined in the early morning when peak secretion should be present (see Chapter 5 Section 4(i)). Although there is some circadian variation in thyroid hormones it is not marked and samples for thyroxine can be taken at any time of day. Thyroxine levels of <50 nmol/l and morning cortisol of <200 nmol/l indicate TSH and ACTH deficiency respectively. Dynamic pituitary function tests (Section 12:2) may be applied to confirm these pituitary deficiencies since direct measurements of basal levels of TSH and ACTH are not instructive due to the large normal variations in the concentrations of these hormones in blood.

Deficiency of the neurohypophyseal hormone, vasopressin, also occurs rarely and the ensuing disease is known as diabetes insipidus. The condition is characterised by the lack of antidiuretic activity which results in frequent micturition and insatiable thirst, and can be confirmed by a water deprivation test (see Section 2). This test is also useful to distinguish vasopressin deficiency from nephrogenic diabetes insipidus which is either a genetic defect or an acquired inability of the renal tubules to respond to vasopressin. A synthetic long-acting analogue of vasopressin, known as desmopressin (DDAVP), has greater antidiuretic potency than vasopressor activity (see Chapter 3 Section 9) and may be used to treat diabetes insipidus patients by intranasal administration.

(ii) *Hyperpituitary Function*

Overproduction and hypersecretion of a number of pituitary hormones occurs usually as a result of a secretory tumour (see [8]); those which most commonly occur are the peptide or protein hormones, ACTH, growth hormone or prolactin. Cushing's syndrome is the name given to the clinical condition induced by excess corticosteroids irrespective of the cause, whereas the term Cushing's disease, the most common form in adults, indicates that the cause is an ACTH-secreting tumour of the pituitary. Cushing's syndrome is indicated by loss of circadian rhythm of blood corticosteroids usually determined by elevated urine steroid excretion throughout the 24 hours; but stress, due to hospitalisation for example, or depression or obesity can result in high steroid levels. Also in Cushing's syndrome, the normal response to insulin induced hypoglycaemia is usually lost. Differentiation between Cushing's disease and other causes of Cushing's syndrome may be difficult and further dynamic tests may be necessary (see Section 12:2). Treatment of Cushing's syndrome may involve pituitary irradiation to remove the tumour or reduction of plasma cortisol by drugs or adrenalectomy.

A growth-hormone-secreting adenoma in a child or adolescent is very rare but can cause gigantism (precocious growth) whereas in an adult (where the bone epiphyses have fused to prevent excessive growth) such an adenoma is more common and causes acromegaly. The latter is characterised by excessive growth of bones and soft tissues in hands, feet, jaws and supraorbital brow ridges. In acromegaly or gigantism blood levels of growth hormone tend to be high and show resistance to normal suppression. The confirmatory dynamic test for this involves a glucose tolerance test (see Section 12:2), which is based on the normal inhibitory effects of glucose on growth hormone secretion. Some acromegalic patients secrete growth hormone in response to TRH but this is not usually used as a diagnostic test. The disease may be treated by surgical ablation or irradiation of the tumour but in many instances can be controlled by dopamine agonist (e.g. bromocriptine) treatment (see below). Dopamine normally increases growth hormone levels in blood (see Chapter 7 Section 3(iv)) but paradoxically such drugs lower hormone levels in acromegaly.

Pituitary tumours secreting prolactin may be prolactinomas or mixed hormone-secreting adenomas (often micro-adenomas). However, there are other causes of raised serum prolactin (hyperprolactinaemia) including normal pregnancy and lactation, stress, sleep, drugs, as well as hypothalamic disorders. Hyperprolactinaemia can lead to galactorrhoea, hypogonadism, amenorrhoea and infertility in women and whilst the condition is less common in men it may lead to impotence, ejaculatory failure or galactorrhoea. The reasons for this relate to the reduced actions of the gonadotrophins on the gonads in the presence of high prolactin and inhibition of the positive feedback of oestrogens which trigger the LH surge (see Chapter 7 Section 4). The condition can only be confirmed by radioimmunoassay measurements of serum prolactin since reliable and conclusive dynamic tests are not available. Hyperprolactinaemia caused by pituitary or hypothalamic disease may be treated with a long-acting dopamine receptor agonist drug, such as bromocriptine (2-bromo-αergocryptine) (see Chapter 7 Section 4(i)).

Vasopressin excess is fairly common in clinical medicine (called syndrome of inappropriate ADH; SIADH) and tends to complicate many diseases of the chest and brain. SIADH results when a reduction in plasma osmolality does not reduce vasopressin secretion and this may cause hyponatraemia (low plasma sodium concentration) due to excess water reabsorption. The condition usually does not require treatment other than water deprivation to reduce the patient's water intake to below his or her urine output. Oxytocin excess or deficiency is not associated with any disease states. For further reading on hyperpituitary function see [1, 4, 8].

12:2 Pituitary Function Tests

The purpose of this section is to provide an indication of the dynamic function tests commonly used in endocrine clinics to assess pituitary hormone secretion. Space does not permit a comprehensive cover of all tests available, nor will details of the individual tests be given, but further references are given at the end of the Chapter. More detailed aspects of the use of the hypothalamic peptide hormones and their analogues will be covered in the following section (see Section 3).

Pituitary function tests are never undertaken unnecessarily by the clinician since a number are potentially dangerous in certain patients and they can also be expensive in both time and materials. Their purpose is to confirm a diagnosis and either assess adequate hormone reserve (stimulation tests) or determine whether suspected hyperfunction is a 'primary' one associated with the peripheral endocrine gland, or a 'secondary' disorder due to hypothalamic-pituitary dysfunction (suppression tests). In practice, the clinician will use his common sense. For example, it is not necessary to check for normal ACTH reserve of a patient with high blood cortisol, or for normal TSH reserve if the thyroid hormone levels are within the normal expected range or for normal gonadotrophin reserve in a woman who has normal menstrual cycles.

ACTH and growth hormone are normally released by stress. Therefore, a stress test is used to assess normal reserve and response of these hormones. An insulin tolerance, hypoglycaemic stress test may be used, but is particularly dangerous and in children and patients with heart disease or epilepsy, alternative less dangerous tests are employed. An intravenous injection of insulin is given, sufficient to cause hypoglycaemia (reduce blood glucose <2 mmol/l from ~ 15 min to 1 hour) and periodic blood samples are taken for glucose and hormone estimation. In order to interpret the results the normal ranges of basal values and responses are required. A delayed response indicates poor reserve and an exaggerated response indicates hormone excess. The GH reserve of small children with stature below their normal 'age growth curve' can be tested by a glucagon, Bovril (meat extract containing amino acids) or exercise stress test and the measurement of plasma GH (a maximal response <20 mU/l usually indicates GH deficiency). The reserve and secretion of hormones which do not significantly respond to stress, i.e. the gonadotrophins (LH and FSH) and thyrotrophin (TSH), are assessed by specific GnRH (LH-RH) and TRH tests. These involve the intravenous injection of pure synthetic, hypophysiotrophic peptide usually at doses of 200 μg (or 500 μg) and 100 μg respectively (see Section 3 for further details). In cases of suspected hypopituitarism, a combined pituitary function test may be given, which effectively entails insulin tolerance, TRH and LH-RH tests run at the same time where blood samples may be taken to assess glucose, T_4, T_3, cortisol, growth hormone, prolactin, TSH, LH and FSH.

Dynamic suppression tests for hormone hypersecretion in common use include a glucose tolerance test (GTT) for acromegaly. After 50 g oral glucose, serum GH is normally suppressed to <1 mU/l well within 2 hours but no suppression, and even a rise, is found in acromegalic patients. When Cushing's syndrome is established in a patient (see Section 12:1) a differential diagnosis is required to determine whether the cause is a pituitary adenoma, an adrenal adenoma or carcinoma, or ectopic production of ACTH. A 48 hour (high dose) dexamethasone suppression test and a metyrapone test may be used.

2 mg dexamethasone every 6 hours normally reduces serum cortisol levels and urine corticosteroids by 50 per cent within 48 hours and a similar response occurs in Cushing's disease, but not usually with primary adrenal tumours or ectopic ACTH syndrome. Metyrapone (750 mg) orally every 4 hours for 24 hours normally blocks 11-hydroxylase activity and cortisol synthesis in the adrenal (see Figure 5.1). Therefore, plasma cortisol is reduced (and steroid precursor metabolites in urine increase) and ACTH secretion increases due to loss of negative feedback (see Chapter 5 Section 5). In Cushing's disease this response is augmented or exaggerated, whilst patients with primary adrenal tumours show no rise in ACTH or steroid excretion due to the presence of normal suppression. An ectopic source of ACTH may be diagnosed from extremely high ACTH levels, not suppressed.

A water deprivation test is used to diagnose diabetes insipidus (see Section 1). It consists of an 8 hour day-time deprivation of water from a weighed patient, who is closely supervised in order to prevent dangerous and extreme dehydration if the vasopressin deficiency is severe (indicated by >3 per cent fall in body weight) and to prevent cheating by a compulsive drinker. Normal patients concentrate their urine (osmolality >600 mOsm/kg) to maintain normal plasma osmolality (at <300 mOsm/kg) and to prevent dehydration. In diabetes insipidus plasma osmolality will rise and urine osmolality remains low. Failure to produce a concentrated urine during water deprivation should be restored by vasopressin or DDAVP. If this does not occur then nephrogenic diabetes insipidus is indicated (see Section 1). For further details on pituitary function tests see [4].

12:3 Clinical Uses of Hypothalamic Peptide Hormones

Following the structural characterisation of TRH, Gn-RH and somatostatin between 1969 and 1972, synthetic preparations became widely available to investigators throughout the world and clinical studies were initiated. Today their use can no longer be considered experimental, but rather they constitute an integral part of the clinical endocrinologist's diagnostic procedure. The clinical value of the synthetic peptides relates to the information obtained by measurement of the circulating levels of pituitary hormones following administration of these releasing hormones. The assessment of pituitary responses, both qualitative and quantitative, has enabled clinicians to determine pituitary function and to define certain parameters of hypothalamic-pituitary disease (see Section 12:2). Each of the three hormones are discussed briefly below and each section presents, in order, the effects of hypothalamic hormones on pituitary hormone secretion in normal subjects, factors that modify the pituitary responses and effects of hypothalamic hormones in pathological conditions.

Several laboratories have reported measurements of TRH and Gn-RH in peripheral plasma, but the levels from different laboratories show considerable variability and have not been demonstrated to vary under physiological or pathological circumstances. These findings are not so surprising since enzymes in the plasma rapidly break down these peptides. Thus it is unlikely that their measurement in peripheral blood will be of any major clinical importance.

(i) *Thyrotrophin Releasing Hormone (TRH)*

The response to a single intravenous injection of TRH, at a dose of around 200 μg, is a rapid rise in the circulating levels of TSH peaking at 15 to 30 min and returning to base line after approximately 90 minutes (see Figure 12.1). Repeated injections of TRH diminish the TSH response due to an elevation of endogenous thyroid hormone which exert an inhibitory feedback on the pituitary TSH response to TRH. In fact administration of either tri-iodothyronine (T_3) or thyroxine (T_4) completely abolishes the TSH response (see Chapter 6 Section 5). As might be expected from these observations hyperthyroidism, whether it be due to autoimmune complications (Graves disease), an autonomous thyroid disorder, or very rarely a TSH secreting pituitary adenoma, is characterised by decreased TSH responses to TRH (Figure 12.1) whereas enhanced responses are obtained in primary hypothyroidism. Secondary (i.e. pituitary or hypothalamic) hypothyroidism may also be investigated using TRH. If the patient is hypothyroid and the TSH response to TRH is not excessive this indicates a secondary cause since secondary hypothyroidism is usually associated with an impaired TSH response. It is interesting to note however that a postulated TSH insufficiency in patients with known hypothalamic disease does not always impair the pituitary TSH responses, which contrasts the impaired gonadotrophin response to Gn-RH after periods of Gn-RH deficiency (see Section 3(ii)). Thus, TRH tests can be of great diagnostic value, distinguishing between primary and secondary hypothyroidism or excluding primary hyperthyroidism. Furthermore, prior to the availability of synthetic TRH, the most useful dynamic test available to exclude primary hyperthyroidism (thyrotoxicosis) was the T_3 (tri-iodothyronine) suppression test which required

Figure 12.1: Prolactin and TSH Responses to an Intravenous Injection of 200 μg Synthetic TRH at time 0 in (a) a normal subject and (b) a patient with hyperthyroidism: note the impaired TSH response but normal prolactin response. (From J. Jenkins with permission.)

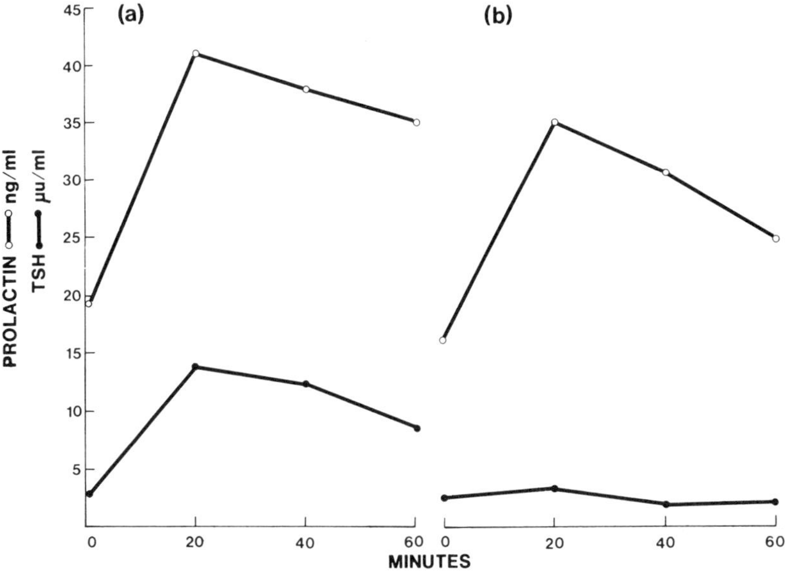

a time consuming study of uptake of radiolabelled iodine by the thyroid gland before and after 7 days treatment with supraphysiological doses of T_3.

TRH is also a potent stimulator of prolactin secretion and the response threshold appears to be similar to that of TSH. Although the physiological significance of this effect is in doubt (see Chapter 7 Section 4(ii)) the monitoring of such responses is of clinical value. For example, the prolactin response to TRH is frequently altered in patients with pituitary disorders and the response may be used to test the integrity of the pituitary secretory reserve of prolactin. Prolactin responses to TRH are greater in women than men due to the higher levels of oestrogens (see Chapter 7 Section 5(iii)). In hyperprolactinaemia, an absent or reduced prolactin response to TRH suggests the presence of a prolactin-secreting tumour, and following treatment the restoration of prolactin responsiveness may be monitored. In addition to its effects on TSH and prolactin secretion, TRH has also been shown to stimulate the release of growth hormone under certain specific conditions including anorexia nervosa, acromegaly, depression and hypothyroidism, and evidence suggests that the GH stimulation is due to a pituitary site of action of TRH.

TRH not only acts on the pituitary but extensive reports in the literature have described its neuropharmacological and behavioural effects in the CNS (see Chapter 6 Section 3(ii) and Chapter 11 Section 7). The pituitary responses to TRH are altered in various psychiatric disorders, such as reduced TSH responses observed in depressed patients and also the paradoxical GH responses already referred to above. For these reasons TRH may be of value in clinical psychoneuroendocrine studies (see [6]) but despite earlier encouraging reports the diagnostic and therapeutic value of TRH in depression and other psychiatric disorders is limited.

A significant number of TRH analogues have been synthesised in an attempt to produce more stable and potent molecules (see Figure 6.3). Investigations with these analogues have indicated differences in receptor potency between pituitary and brain sites. Only very few TRH analogues with modifications in the His residue show increased activity on TSH and prolactin secretion (e.g. [3^{im}Me]His TRH). Modifications of pyroGlu or ProNH_2 generally reduce pituitary potency but a number of such analogues (e.g. pyroGly-His-DmpNH_2 (see Figure 6.3)) show increased potency in a number of pharmacological tests in the CNS (see [1, 4, 12] in Chapter 6). The potential clinical value of these and other TRH analogues are at present undergoing assessment.

(ii) *Gonadotrophin Releasing Hormone (Gn-RH)*

Administration of Gn-RH releases both LH and FSH, the LH response being considerably greater than the FSH response with a peak occurring between 30 and 60 minutes post injection (see Figure 12.2). However, the onset of the FSH response is delayed and there is a much slower return to baseline levels due in part to the fact that the metabolic clearance rate is about one quarter that of LH. Steroids can markedly alter the quantitative LH and FSH responses: during the menstrual cycle greater responses are observed mid-cycle when the oestrogen levels are relatively high (see Chapter 8 Section 3), and steroid treatment can either enhance or depress gonadotrophin responses to Gn-RH depending on dose and sex. Generally speaking the release of LH in response to exogenous Gn-RH is greatest under those conditions in

Figure 12.2: (a) LH and (b) FSH Responses to an Intravenous Injection of 100 μg Synthetic Gn-RH in a Normal Subject at Two Different Stages of the Menstrual Cycle: note the enhanced responses observed mid-cycle around the time of ovulation.

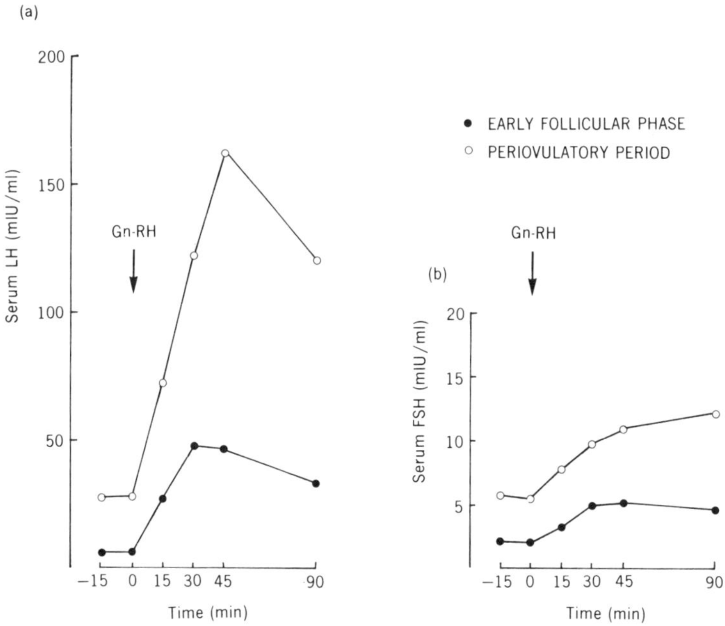

which endogenous Gn-RH is being secreted and sensitises the pituitary. Hence, in prepubertal children, when Gn-RH secretion is probably very low, the LH response is minimal and is, in fact, less pronounced than the FSH response. Similar prepubertal-type responses are observed in women with anorexia nervosa and it has been shown that with increasing weight loss there is a progressive decrease in pituitary responsiveness to Gn-RH. Conversely, after the menopause when ovarian steroid secretion has declined the LH response to Gn-RH is increased. In some instances gonadotrophin responses to exogenous Gn-RH have been useful in distinguishing between pituitary and hypothalamic disorders although the marked variability in responses associated with changes in circulating levels of gonadal steroids and

Figure 12.3: (a) The Structure of Gn-RH. (b) Some examples of superactive analogues and antagonists of Gn-RH. Note that in the first example of each group the full molecular structure is given but subsequent examples only indicate the substitutions.

(a) GONADOTROPHIN RELEASING HORMONE

1 2 3 4 5 6 7 8 9 10
(pyro) Glu-His-Trp-Ser-Tyr-Gly-Leu-Arg-Pro-Gly-NH_2

(b) EXAMPLES OF SUPERACTIVE ANALOGUES

(i) (pyro) Glu-His-Trp-Ser-Tyr-D-Ala[6]-Leu-Arg-Pro[9]-NH-CH_2-CH_3

D-Ala^6 Gn-RH (1-9) ethylamide

(ii) D-Ser$(Bu^t)^6$ Gn-RH (1-9) ethylamide (buserelin)

(iii) D-Leu^6 Gn-RH (1-9) ethylamide

(c) EXAMPLES OF Gn-RH ANATAGONISTS

(i) (pyro) Glu-D-Phe[2]-D-Phe[3]-Ser-Tyr-D-Phe[6]-Leu-Arg-Pro-Gly-NH_2

D-Phe^2, D-Phe^3, D-Phe^6 Gn-RH

(ii) D-Phe^2, D-Trp^6 GnRH

(iii) D-Phe^2, D-Trp^3, D-Phe^6 Gn-RH

(iv) D-(pyro) Glu-D-Phe-D-Trp-Ser-Tyr-D-Lys-Leu-Arg-Pro-Gly-NH_2

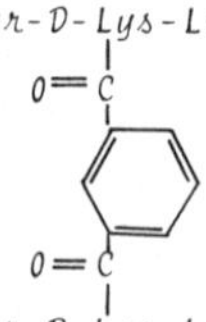

D-(pyro) Glu-D-Phe-D-Trp-Ser-Tyr-D-Lys-Leu-Arg-Pro-Gly-NH_2

Isophthaloyl dimer of D-(pyro) Glu^1, D-Phe^2, D-Trp^3, D-Lys^6 Gn-RH rates of endogenous Gn-RH secretion poses severe limitations on elucidating specific pathophysiological conditions.

Although Gn-RH tests are not diagnostically very valuable, synthetic Gn-RH and structurally related analogues have great clinical potential both as contraceptives and fertility agents. Well over 500 Gn-RH analogues have been synthesised and stable superactive agonists and antagonists have been produced. Early results showed that small fragments of the decapeptide were not biologically active and that in general amino acids in position 1 and from 4 to 10 (see Figure 12.3) appeared to be only involved with binding to Gn-RH receptors and in exerting confirmational effects. However, histidine and tryptophan in positions 2 and 3 appear to take part in the chemical mechanisms activating adenylate cyclase as well as providing receptor binding capacity.

Superactive analogues have been synthesised by substitutions of the amino acids in the 6 and 10 positions of the Gn-RH molecule (see Figure 12.3) and these analogues are active when administered by a variety of peripheral routes including intravenously, subcutaneously, orally, rectally and intransnasally. The substitution of glycine in position 6 by D-amino acids, results in analogues up to 15 times more active than Gn-RH for gonadotrophin release. Similarly if the glycinamide residue in position 10 is replaced by simple alkylamine groups, again gonadotrophin releasing activity is increased. However substitutions in both positions (6 and 10) result in analogues which can release 50–60 times more LH and 15 times as much FSH as the same dose of the native Gn-RH. It has been reported that these analogues are more resistant to peptidase activity in rat serum and hypothalamic and pituitary homogenates, although it is more likely that their prolonged activity is due to their binding capacity and protracted time course of binding to their pituitary receptors.

The histidine and tryptophan (2 and 3) residues can be altered in a number of ways to give analogues with almost no gonadotrophin releasing activity but which retain affinity for Gn-RH receptors. Substitution of the histidine with D-phenylalanine and the glycine in position 6 with either D-phenyalanine or D-lysine creates a relatively long lasting, effective antagonist of Gn-RH. More potent antagonists are produced by additionally substituting the tryptophan in position 3 with either D-tryptophan or D-phenylalanine (see Figure 12.3). More recently dimers and trimers of Gn-RH antagonists have been synthesised, some of which have enhanced inhibitory activity compared with their parent monomers. It is likely that both stimulatory and inhibitory analogues may exert their action on the same pituitary receptors as the native Gn-RH since both superactive analogues and potent antagonists can compete with Gn-RH for its receptors, and both analogues displace ^{125}I-labelled Gn-RH more strongly than the parent hormone.

One might predict that superactive analogues could be potentially valuable in stimulating ovulation. However, paradoxically long-term treatment with pharmacological doses of superactive analogues leads to impaired reproductive functions. Hence daily administration of Gn-RH agonists during the menstrual cycle suppresses ovulation and induces luteolysis in both humans and sub-human primates. In the rat Gn-RH analogues have also been shown to block implantation and terminate gestation, and it has been suggested that such inhibitory effects on luteal function may result from the sustained secretion of LH and the ability of LH to decrease the availability of its own receptors (down regulation) on the corpus luteum (see Chapter 8 Section 2). Recent evidence has suggested that ovulation can only be induced in humans by pulsatile infusions of Gn-RH over a period of several days comparable to the experimental observations on monkeys (see Chapter 9 Section 1). Thus another reason why potent long-acting Gn-RH analogues do not appear to be suitable for the treatment of infertility is that they do not provide the correct signal for the pituitary gland to respond in the appropriate manner for induction of ovulation.

In contrast there is a potential clinical use for Gn-RH antagonist analogues in suppressing ovulation. In several animal species Gn-RH antagonists have been shown to inhibit ovulation although work on primates has been less convincing. Nevertheless it has been shown that some potent antagonists can suppress LH secretion both in monkeys and man. The development of potent Gn-RH analogues has far-reaching clinical applications both as potential pre-coital and post-coital contraceptives. However, at the present time therapeutic regimens are largely empirical and, in view of the antifertility effects of long-term treatment with superactive analogues, more work is required to establish their precise mode of action and in devising safe clinical protocols. For further reading see [3, 9, 10].

(iii) *Somatostatin*

Somatostatin is a powerful inhibitor of growth hormone secretion and it can suppress the spontaneous sleep-associated surges of GH secretion and inhibit the increases in GH which occur in response to such stimuli as insulin-induced hypoglycaemia, exercise and arginine infusions (see Chapter 7 Section 2(ii)). In addition somatostatin is also an inhibitor of TSH secretion in both normal and hypothyroid subjects and it can block the TSH response to TRH administration, but at the present time, there is insufficient evidence to establish a physiological role of somatostatin in TSH release. The response to intravenous infusions of somatostatin in acromegalic patients is a rapid lowering of GH levels to normal baseline levels with a rebound secretion immediately upon cessation of the infusion. In normal subjects somatostatin does not affect the secretion of LH, FSH, ACTH and prolactin although in some acromegalics somatostatin is capable of suppressing prolactin secretion. Other abnormal responses to somatostatin have also been observed in acromegaly and various endocrine disorders, although, in contrast to TRH, such responses have not proved to be of great diagnostic value.

Apart from its inhibition of GH and TSH, somatostatin has widespread effects on both the endocrine and exocrine secretions in the gut. It suppresses the secretion of insulin and glucagon and decreases the release of several gut hormones including secretin, gastrin and cholecystokinin. These observations, linked with the immunohistochemical evidence of a widespread distribution of somatostatin in the gut, support the view that this so-called hypothalamic release-inhibiting hormone plays a role, not only in anterior pituitary function, but also in secretory functions of the pancreas, duodenum, stomach and possibly other tissues.

Somatostatin itself (see Figure 7.2) is of little therapeutic value due to its multiple actions and its short biological half life (~4 min). Nevertheless, analogues of somatostatin have been synthesised which show prolonged activity and the ability to inhibit selectively certain hormonal secretions. For example (D-Trp^{8}, D-Cys^{14}) somatostatin has a ratio of 22:1 for the selective inhibition of glucagon over insulin and 100:1 for that of GH over insulin. A further analogue, des-Asn (D-Trp^{8}, D-Ser^{13}) somatostatin, is about twenty times more potent in inhibiting insulin secretion compared with the parent compound and is less potent in inhibiting glucagon and GH. The synthesis of long-acting analogues which possess differential potencies on the various physiological secretions affected by somatostatin may thus have important clinical implications in the treatment of acromegaly, juvenile diabetes, peptic ulcers and other diseases and research in this area looks promising. For further information see [10].

(iv) *Other Biologically Active Peptides*

A large number of biologically active peptides are present in relatively high concentrations in both the hypothalamus and pituitary and many animal studies have indicated that pituitary secretions may be affected by a variety of peptides other than the so-called hypothalamic releasing and inhibiting hormones (see Chapter 10 Section 5).

Clinical studies on the role of such neuropeptides have concentrated on the opioid-like peptides, endorphins and enkephalins, since their administration causes significant endocrine and metabolic effects. For example a long-acting analogue of Met-enkephalin stimulates the release of GH, prolactin and TSH whilst inhibiting the secretion of LH, FSH, ACTH and β-lipotrophin. Naloxone, a non-peptide opiate antagonist, has also been used to investigate the possible involvement of endogenous opioid peptides in the control of anterior pituitary secretions. However, since different opioid receptors have been defined which show different sensitivities to the antagonistic action of naloxone, the interpretation of naloxone 'effects' are complex, and it is further possible that there may be opiate effects at sites which are completely naloxone insensitive. In addition there is clinical evidence that endogenous opioid peptides may modulate either the responsiveness of the pituitary to hypothalamic releasing hormones or may potentiate or inhibit stimulated-release (as opposed to tonic release) by a hypothalamic site of action. Clinical studies pertaining to the involvement of biologically active peptides other than TRH, somatostatin and Gn-RH in the control of anterior pituitary secretions are clearly in their infancy and to date there is no evidence to indicate any diagnostic or therapeutic value in endocrine disorders. For further information on the development of biologically active peptides as drugs see [5]. For further reading on the clinical use of non-peptide drugs which are known to modify the secretion of anterior pituitary hormones see [2, 8] and D. T. Krieger & J. C. Hughes (eds) (1980) *Neuroendocrinology – a Hospital Practice Book* (Sinauer Assoc. Inc., Sunderland, MA).

12:4 Future Developments

The major technical improvements in peptide chemistry and synthesis, coupled with recent development of some potent peptide analogues, provides a glimmer of success in designing potentially useful drugs.

There are three general requirements for producing clinically effective agonist and antagonist analogues: (1) reduction of metabolism, (2) maintenance of receptor specificity, and (3) increased membrane transport for oral administration and penetration into the brain (see Chapter 11 Section 4). Analogues of Gn-RH have been shown to fulfil these requirements, and there is little doubt of their potential clinical significance as fertility agents.

To date, the search for active peptide agonists has been more successful than for peptide antagonists, and the existence of heterogeneous populations of receptors had aided the development of agonists which show selective activity. Such receptor specificity has already been described for analogues of somatostatin and a further example which may be pertinent to neuroendocrine systems is an analogue of TRH (p-Glu-His-Pro-NH $(CH_2)_6$ NH_3) which is fifteen times more potent in stimulating the synthesis of prolactin than in stimulating the release of prolactin. At the present time specific receptor antagonists are available only for the opioid peptides, Gn-RH, Substance P, angiotensin and vasopressin and studies with these antagonists strongly indicate the importance of developing a range of antagonists which show specificity in heterogeneous populations of receptors.

The potential for the development of non-peptide agonist and antagonists that can alter peptidergic function is apparent. The use of the opioid agonist morphine, for instance, and antagonists such as naloxone were not only essential in the discovery of endogenous opiate-like peptides and subsequent testing of their possible physiological roles, but are also clinically important. A further approach to modulating the action of centrally active peptides is through drugs which affect the synthesis or metabolism of peptides without acting on specific receptors. The development of inhibitors to the enzyme which converts angiotensin I into the active angiontensin II provides a good example of indirect modulation of a peptide system, and preliminary clinical trials have already indicated that such inhibitors may be an effective anti-hypertensive agent. In summary, recent research has opened up new experimental avenues in which to explore and develop new drugs that may alter endogenous biologically active peptides. Areas of drug design include the synthesis of peptide or non-peptide agonists and antagonists and the production of indirectly acting drugs which can alter various enzyme activities.

In the present section emphasis has been placed on analogues which may alter hypothalamic-pituitary function, but it should be pointed out that the role of peptides in neuroendocrine function constitute a small part of the entire experimental field of biologically active peptides. Also peptides which may alter the secretion of pituitary hormones may have more profound effects on other physiological systems – again such an example are the opioid peptides which may be more clinically relevant in analgesia and psychiatric disorders than in neuroendocrine regulation.

References

1. Besser, G. M. & Mortimer, C. H. (1976) Clinical Neuroendocrinology, in L. Martini, W. F. Ganong (eds) *Frontiers in Neuroendocrinology*, Vol. 4. (Raven Press, New York), pp. 227-54.
2. Del Pozo, E. & Lancranjan, I. (1978) Clinical Use of Drugs Modifying the Release of Anterior Pituitary Hormones, in W. F. Ganong & L. Martini (eds) *Frontiers in Neuroendocrinology*. Vol. 5. (Raven Press, New York), pp. 207-47.
3. Frohman, L. A. (1978) Newer Understanding of Human Hypothalamic Pituitary Disease Obtained through the use of Synthetic Hypothalamic Hormones, in S. Reichlin, R. J. Baldessarini & J. B. Martin (eds) *The Hypothalamus* (Raven Press, New York), pp. 387-413.
4. Hall, R., Anderson, J., Smart, G. A. & Besser, M. (1980 III Ed) *Fundamentals of Clinical Endocrinology* (Pitman Medical).
5. Hughes, J. (1980) Biologically Active Peptides: Prospects for Drug Development. *Philosophical Transactions of the Royal Society, London*, Series B, **290**: 387-94.
6. Loosen, P. T. & Prange, A. J. (1980) TRH: a Useful Tool in Psychoneuroendocrine Investigation. *Psychoneuroendocrinology*. **5**: 63-80.
7. Muller, E. E. (ed.) (1980) *Neuroactive Drugs in Endocrinology: Physiologic, Diagnostic and Therapeutic Applications* (Elsevier/North Holland).
8. Post, K. D., Jackson, I. M. D. & Reichlin, S. (eds) (1980) *The Pituitary Adenoma* (Plenum Publ. Corp., New York).
9. Schally, A. V., Arimura, A. & Coy, D. H. (1980) Recent Approaches to Fertility Control Based on Derivatives of LH-RH. *Vitamins and Hormones*, **38**: 258-324.
10. Schally, A. V., Coy, D. H. & Meyers, C. A. (1978) Hypothalamic Regulatory Hormones. *Annual Review of Biochemistry*, **47**: 89-128.

(See also General Book References 14 and 16.)

GENERAL BOOK REFERENCES

1. Barker, J. L. & Smith, T. G. (eds) (1980) *The Role of Peptides in Neuronal Function* (Marcel Dekker Inc., New York).
2. Barrington, E. J. W. (ed.) (1979) *Hormones and Evolution* (Academic Press, New York).
3. Bloom, S. R. & Polak, J. M. (eds) (1981) *Gut Hormones* (Churchill Livingstone, Edinburgh).
4. Collu, R., Barbeau, A., Ducharme, J. R. & Rochefort, J.-G. (eds) (1979) *Central Nervous System Effects of Hypothalamic Hormones and Other Peptides* (Raven Press, New York).
5. Costa, E. & Trabucchi, M. (eds) (1980) *Neural Peptides and Neuronal Communication* (Raven Press, New York).
6. Fuxe, K., Gustafsson, J.-A. & Wetterberg, L. (eds) (1981) *Steroid Hormone Regulation of the Brain* (Pergamon Press).
7. Ganong, W. F. & Martini, L. (various volumes, 1969-1980) *Frontiers in Neuroendocrinology* (Oxford University Press).
8. Gupta, D. & Voelter, W. (eds) (1978) *Hypothalamic Hormones – Chemistry, Physiology and Clinical Applications* (Verlag Chemie Weinheim, New York).
9. Heller, H. & Lederis, K. (eds) (1971) *Subcellular Organisation and Function in Endocrine Tissues* (Cambridge University Press).
10. Hrdina, P. D. & Singhal, R. L. (eds) (1982) *Neuroendocrine Regulation and Altered Behaviour* (Croom Helm, London).
11. James, V. H. T. & Pasqualini, J. R. (eds) (1980) *Hormonal Steroids* (Pergamon Press).
12. Jeffcoate, S. L. & Hutchinson, J. S. M. (eds) (1978) *The Endocrine Hypothalamus* (Academic Press, New York).
13. Jutisz, M. & McKerns, K. W. (eds) (1980) *Synthesis and Release of Adenohypophyseal Hormones* (Plenum Publ. Corp. New York).
14. Krieger, D. T. & Hughes, J. C. (eds) (1980) *Neuroendocrinology – a Hospital Practice Book* (Sinauer Assoc. Inc., Sunderland, Massachusetts).
15. Maddrell, S. H. P. & Nordman, J. J. (1979) *Neurosecretion. Tertiary Level Biology* (Blackie, Glasgow).

16. Martini, L. & Besser, H. M. (1977) *Clinical Neuroendocrinology* (Academic Press, New York).
17. Meites, J., Donovan, B. T. & McCann, S. M. (1978) *Pioneers in Neuroendocrinology I and II* (Plenum Publ. Corp., New York).
18. Millar, R. P. (ed.) (1981) *Neuropeptides: Biochemical and Physiological Studies* (Churchill Livingstone, Edinburgh).
19. Morgane, P. J. & Panksepp, J. (1979, 1980) *Handbook of the Hypothalamus*. Vol. I Anatomy, Vol. II Physiology (Marcel Dekker Inc., New York).
20. Reichlin, S. Baldessarini, R. J. & Martin, J. B. (eds) (1978) *The Hypothalamus* (Raven Press, New York).

INDEX

Note: page numbers in italics denote figures.